Construction Partnering & Integrated Teamworking

Mike Thomas Ltd

Facilitation, Procurement & Value Management

Since setting up Mike Thomas Ltd in December 1997, Gill and Mike Thomas have facilitated more than four hundred project and strategic team workshops including partnering, value and risk management, post-project reviews, team-building and best practice training.

Mike Thomas qualified as a quantity surveyor in 1972 whilst working in private practice. From 1974 to 1997 he worked within a major client organisation as quantity surveyor, project manager, research and development manager and, finally, strategic construction procurement manager. During his time in R&D and procurement, Mike managed the introduction and implementation of many innovations and initiatives including Value Management, alternative construction methods and Partnering.

Gill Thomas obtained her BA at London University in 1972 and Certificate of Education at Southampton University in 1973. She taught French and Latin to senior school pupils and French to adult education students before sons Philip and Andrew were born in 1977 and 1978. Gill returned to secondary school teaching in 1985 and was seconded to train as a Teacher of Hearing Impaired Children. Before leaving teaching in 1998, she learnt sign language, taught French to hearing-impaired children to GCSE level, mentored newly-qualified teachers and managed the hearing-impaired department of a 1200-pupil school.

Gill and Mike have been married since 17 August 1974.

39 Atlantic Close, Ocean Village, Southampton, SO14 3TB
Tel: 023 8023 1169, Email: mikethomasltd@btconnect.com

Construction Partnering & Integrated Teamworking

Gill Thomas and Mike Thomas

Mike Thomas Ltd, Partnering Facilitators
Ocean Village, Southampton

Editorial offices:
Blackwell Publishing Ltd, 9600 Garsington Road, Oxford OX4 2DQ, UK
Tel: +44 (0)1865 776868
Blackwell Publishing Inc., 350 Main Street, Malden, MA 02148-5020, USA
Tel: +1 781 388 8250
Blackwell Publishing Asia Pty Ltd, 550 Swanston Street, Carlton, Victoria 3053, Australia
Tel: +61 (0)3 8359 1011

First published 2005 by Blackwell Publishing Ltd

Library of Congress Cataloging-in-Publication Data
Thomas, Gill.
Construction partnering and integrated teamworking / Gill Thomas and Mike Thomas.
p. cm.
Includes bibliographical references and index.
ISBN-10: 1-4051-3556-5 (pbk. : alk. paper)
ISBN-13: 978-1-4051-3556-6 (pbk. : alk. paper)
1. Construction industry–Management. 2. Teams in the workplace. 3. Strategic alliances (Business). 4. Partnership. I. Thomas, Mike. II. Title.
HD9715.A2T487 2005
624'.068'4–dc22
2005004147

ISBN-10: 1-4051-3556-5
ISBN-13: 978-14051-3556-6

A catalogue record for this title is available from the British Library
Set in 10/13pt Palatino
by SPI Publisher Services, Pondicherry, India
Printed and bound in India
by Replika Press Pvt Ltd, Kundli

The publisher's policy is to use permanent paper from mills that operate a sustainable forestry policy, and which has been manufactured from pulp processed using acid-free and elementary chlorine-free practices. Furthermore, the publisher ensures that the text paper and cover board used have met acceptable environmental accreditation standards.

For further information on Blackwell Publishing, visit our website:
www.blackwellpublishing.com

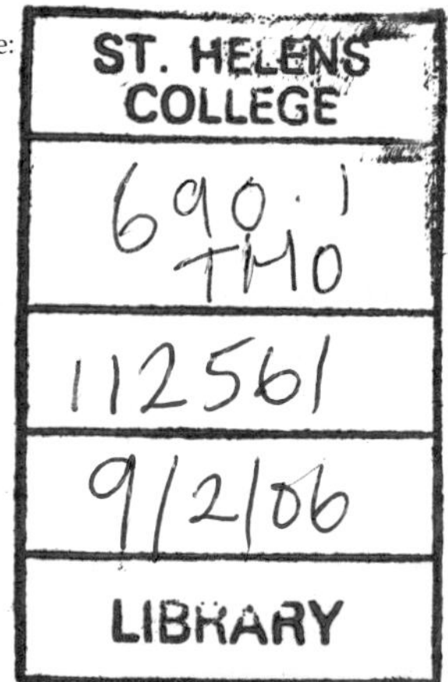

Contents

Preface

Our aim in writing this book is to assist managers of small, medium and large organisations who have been tasked with implementing partnering and integrated teamworking in the public or private sector.

The book brings together the specific processes and tools that we have used and developed for a wide range of teams since first facilitating integrated teamworking through value management since 1990 and partnering since 1993. We have included many examples of best practice from our own experience and are grateful to the teams who have allowed us to use their names and projects in this way. We have included sample agendas for workshops and models for other processes which we encourage you to adopt and adapt to meet the specific needs of your own partnering and integrated teams.

We acknowledge that the cultural shift from lowest price purchasing to best value procurement through partnering and integrated teamworking is challenging. Therefore, our objective is to provide practical help in an accessible format for those actively and peripherally involved in construction. A further objective is that the chapters are concise yet comprehensive. The topics that we have addressed are relevant to project and strategic partnering relationships, term contracts and frameworks and we are sure that this book will be useful not only to specialists, constructors, consultants and clients in both public and private sectors but also to interested parties including auditors, tenants, facilities managers, board members and elected members involved in the delivery of best value.

This book is not intended to be read through at one sitting or consecutively from front cover to back. We anticipate that readers will cherry pick chapters and topics as appropriate to their needs. Thus, some principles are repeated through the book in order to ensure that each chapter can stand alone.

In various sections of the book we have commented that, for example, one process may be more or less suitable than another in a particular situation or that a workshop should be held at a specific point in the partnering programme. These are our opinions based on our experience of working with various integrated teams but without any knowledge of your specific team or project objectives. Therefore, please use our views to guide your thinking and to help form your processes. Partnering and integrated teamworking is not an exact science – people, projects and teams differ in their chemistry and their needs. We know that all of the processes in this book can be adapted to fit your requirements because we use these processes and adapt them on every workshop that we facilitate, to suit the specific needs of the project and team.

We would like to thank the many teams for whom we have facilitated workshops – the innumerable colleagues who have been fun to work with and who have themselves been good team players. We would especially like to thank Roger Harris and the team at Construction Study Centre for their proactive support, friendship and challenges. Roger regularly rings us with opportunities to write new courses in response to requests from his customers and this keeps us focused on continuous improvement.

Gill and Mike Thomas
Southampton

1 The Culture Change

In his introductory statement to *Accelerating change* (Egan, 2002), Sir John Egan said that 'Integrated team working is key. Integrated teams deliver greater process efficiency and by working together over time can help drive out the old style adversarial culture, and provide safer projects using a qualified, trained workforce. It is self evident that teams that only construct one project learn on the job at the client's expense and hence will never be as efficient, safe, productive or profitable as those that work repeatedly on similar projects. I want to see expert teams coming together to deliver world class products, based on understanding client needs.'

This identification of the need for a cultural shift in the construction industry was set out in *Constructing the team* (Latham, 1994) and was driven forward in *Rethinking construction* (Egan, 1998) and in *Delivering better services for citizens* (Byatt, 2001). They all proposed a move away from tendering solely on the basis of lowest price to a value-based selection process including a shifting of emphasis from initial purchase costs and short-term savings to whole-life costs and longer-term objectives to ensure overall best value.

Partnering and integrated teamworking affords a way of achieving better value in whatever way this is defined by the client. However, many organisations are still reluctant to embark on the partnering and integrated teamworking route or are failing to apply a structured approach to lead to major value enhancements in timeliness, better quality and lower costs. Our emphasis on a structured approach is deliberate. In our experience, the greatest value enhancements have accrued to those organisations that select their teams and operate their arrangements in a structured way – planning training, workshops and social events months or even years ahead so that diaries can be committed. Partnering and integrated teamworking, like any other long term relationship, requires a long term commitment.

This book will outline the steps and techniques we have used and developed over the past ten years to introduce and implement successful partnering and integrated teamworking within organisations and project teams in the public and private sectors with clients, consultants, constructors and specialists.

First, the team needs to decide whether it is partnering or whether the separate organisations are forming a partnership. We have noticed that many individuals and organisations use the terms *partnering* and *partnership* interchangeably, frequently using *partnership* in a loose way to describe cooperative working and long term contracts. This is becoming regular practice and it may be that the law will recognise this in due course. However, our understanding is that, as the law stands at present, a partnership is a legal entity in which each of the organisations or individuals that holds itself out as being a partner, is jointly and severally responsible for the debts and obligations of all other partners. This book is focusing on partnering and integrated teamworking as we define below.

The definition of partnering that we propose is a development of a definition first put forward in *Trusting the team* (Bennett & Jayes, 1995). Our definition is, 'an integrated teamworking approach to achieve better value for all partners by reducing duplication and waste of resources, based on mutual objectives, a robust approach to issue resolution and a proactive approach to measurable continuous improvement.'

We see integrated teamworking as a tool in support of the partnering approach but one that could be applicable to all construction projects, not only those with formal or informal partnering arrangements. Our definition of integrated teamworking is taken from the Integration Toolkit published by the Strategic Forum for Construction (http://www.strategicforum.org.uk/sfctoolkit2/home/home.html) 'a single team focused on a common set of goals and objectives delivering benefit for all concerned.'

Perceptions and behaviours across the industry have changed considerably since the early 1990s. The concept of a formal construction contract in which the various members of the team are contracted to trust each other might have seemed like an alien concept to most of the industry fifteen years ago. An increasing proportion of directors, managers and staff from all organisations involved in construction – clients, consultants, constructors, specialists and other interested parties – understand the business case for working collaboratively and the performance of the industry is improving year-on-year, as demonstrated by the construction industry key performance indicators (Constructing Excellence, 2004).

As a result of the changing attitudes and perceptions, there may be a need for partnering training within individual organisations. The need for such training will depend to a great extent on the current culture of the organisations:

- ❑ have they been working collaboratively for some years?
- ❑ does the relationship clearly exhibit all three key features of partnering (mutual objectives, issue resolution and continuous improvement)?
- ❑ do the individual team members understand the value criteria of their own organisations? For example, is defect-free completion worth anything? If so, is it 0.1%. 1.0% or 10.0% of the capital contract value?
- ❑ are there some members in the organisation who, despite an attempt to introduce a partnering culture, are adversarial in their nature and working practices?

Support from a partnering trainer/facilitator will enable the team members to bring their preconceptions into the open within the safe environment of training workshops. It is important that concerns and fears should be aired, assessed and addressed by management and colleagues before embarking on a programme of partnering and integrated teamworking. Dealing with issues in a non-confrontational way, showing respect for each other's views and continually seeking to improve, will help the team to gel and pull in the same direction. Team members will learn to recognise non-partnering behaviour and language and the negative impact these have on the delivery of added value.

Most people are conditioned to oppose change if it is seen as a threat and not as an opportunity. Management must handle the change to partnering and integrated teamworking sensitively if the team is to develop a cooperative culture which delivers better value, in place of an adversarial culture targeted at driving lowest price. Feedback must be sought at all stages from team members, considered and acted upon, to maximise the benefits of integrated teamworking.

The Construction Industry Council highlights the effort needed to maximise the benefits of partnering and integrated teamworking but also underlines the importance of having fun as a team. 'This is where cooperative networks start to form and are shaped so that all members of the team succeed in both their personal and corporate objectives. The aim is to get the team working creatively, cooperatively and even more for them to have fun as a team. Energy and effort put in here will generate creative thinking, understanding and innovative working that will later benefit the team and the project' (Construction Industry Council, 2002).

Because partnering and integrated teamworking requires considerable effort and resource in the early stages, organisations may question the need for partnering and may wish to tender on price as they have always done. However, price-only tendering sets up conflicting objectives within the project team. A key project that is delivered on price yet, through a lack of mutual understanding, misses other client objectives such as timely delivery and fitness for purpose, may reduce value to the client. Partnering and integrated teamworking enables all team members to align their objectives, focusing on the client's objectives whilst identifying and meeting the objectives of all other organisations.

The added value provided through partnering and integrated teamworking will require a clear business case if it is to convince directors who may, themselves, be rewarded by standing orders or company rules that are based on a lowest price strategy. We have worked with project teams who have identified benefits greater than the 10% of total project costs identified in *Trusting the team* (Bennett & Jayes, 1995). Those who are committed to implementing partnering and integrated teamworking must clearly demonstrate the added value of this approach to directors and auditors by quantifying added value from their own experiences or from nationally published case studies.

Once a partnering route is chosen, the integrated team should be selected on the basis of a weighted matrix of price and other value criteria. The selection process should not be a shortlist to pass a quality hurdle, followed by a tender fight to appoint on lowest price. In our opinion, this is only an extension of an approved list and evidence of a sustained lowest price culture. The industry needs to rethink the whole selection process. The team should be brought together as early as possible in order that all share a common understanding of each other's value criteria and the ways in which they are to be delivered. At this early stage, all organisations and team members will have the opportunity to input their own expertise and suggestions, creating a climate for innovation and the delivery of better value for all.

The initial partnering workshop brings the team members together to define their mutual objectives, set up processes for managing the resolution of issues and address opportunities for continuous improvement. Depending on the team's needs, this workshop could be paired with workshops on value and risk management. These workshops will all assist in building the integrated partnering team ethos as well as defining and refining the scheme.

Following early workshops there may be a need to involve specialist sections of the team (task groups) to address further specific topics. The results from the task groups should be fed back to the team through the core group and the partnering champions. The effectiveness of the core group or partnering champions is critical to the success of the relationship. Good communication is key. All team members need to understand their interdependency. If everybody understands each other's roles and responsibilities and can trust each other to do what they say they will do, there should be a significant reduction in wasted resource and added value for all.

During the remainder of the project, the team should meet on a regular basis in continuous improvement workshops which may be targeted at specific areas of the project. These workshops may also afford an opportunity to develop the team through non-project team-focused exercises and social events.

After handover, the team should meet again for a post-project review to celebrate the success of the integrated team and the project, close out any remaining issues, agree and report on KPIs and take forward the successes and opportunities to their next projects. When learning is captured and applied to future projects, all members of the integrated team will benefit from the learning curve and all organisations and individuals will obtain increasingly better value.

To assist the industry to achieve the key targets of *Accelerating change*, the Strategic Forum for Construction launched an Integration Toolkit (Strategic Forum for Construction, 2003). This includes a maturity assessment grid which identifies typical behaviours in key areas of integration and cultural change. For example, under the heading 'awareness', the maturity assessment grid identifies three mindsets:

> Historic – We believe that the industry is made up of individual organisations who are only interested in their own activities
> Transitional – We realise that we can perform better if we understand how those close to us up and down the tiers of the chain are involved
> Aspirational – We understand that the whole industry is interconnected and that most of what we and others do affects each other's performance.
> (www.strategicforum.org.uk/sfctoolkit2/home/home.html)

We believe that there is a drive within the industry to change to a value-based culture. This shift will take time and there will be considerable challenges to individuals and organisations implementing change programmes. However, we have seen the benefits demonstrated in public and private sectors by organisations who have been prepared to commit time, energy and resource to making this work through successful partnering and integrated teamworking.

2 Identifying the Organisation's Value Criteria

No partnering or integrated team can add value to a project unless they understand the specific value criteria for the project – the criteria that will determine whether the project has been successful.

Before it is possible to achieve joint understanding of the value criteria at project team level, each organisation within the integrated team must understand their own organisation's value criteria – determining whether the investment in the project and the integrated team is delivering value to the specific organisation.

There may be a feeling within some organisations that there is little need to develop a specific set of value criteria for each separate project or relationship as each client will have an essentially similar set – for example, price, time and quality. Similarly, the feeling goes, every constructor will have a standard set – for example, overhead, profit and cashflow. But further organisation-specific analysis is required if the criteria are to be set up appropriate to the specific needs of the client, constructor or other team member. Clues to the value criteria for most organisations can be found in their corporate mission statements but developing appropriate and specific criteria (rather than making assumptions on essentially similar criteria) will require a more detailed focus, requiring the organisation to consider issues such as sustainability, whole life costs and safety.

As an example of how organisational value criteria impact on projects, let us take a hypothetical example of two new retail outlets. Organisation A prides itself on swift and efficient service whilst Organisation B has an enviable track record of lowest prices on the commodities it procures and sells. Organisation A is prepared to fund 10% extra capital on the project to halve construction times whilst

Organisation B is prepared to take extra time to save 10% of the project capital cost. As a result of the differing corporate value criteria, the team working for Organisation A will have a completely different set of project or team value criteria from Organisation B's team.

Value criteria are important because they form the basis for many of the measures and processes of partnering and integrated teamworking. Without understanding the value criteria of their own organisations, team members will be less able to make an appropriate value judgement on the suitability of their potential partner organisations. If the organisational values are not aligned then it is likely that the organisations will pull in different directions and neither will obtain better value.

A clear set of organisational value criteria will:

- ❑ help clients and their suppliers (consultants, constructors and specialists) determine their procurement or bidding strategies – for example, whether to partner or not
- ❑ form the basis for the client's structured selection process and the supplier's bid tactics
- ❑ enable partnering organisations to share their value criteria and selfish (corporate) objectives with their partners – the basis for developing a partnering charter of mutual objectives
- ❑ form the basis for key performance indicators
- ❑ enable the team to structure and quantify feedback on performance (through the KPI process and measurement of continuous improvement), proving the benefits of the partnering and integrated teamworking approach and demonstrating added value.

Thus it can be seen that early clarification of value criteria is critical to the success of a relationship, whether for a client, consultant, contractor, specialist or other interested party.

The value criteria for each organisation may be included in the lists set out below. We have summarised these from selfish objectives expressed in initial partnering workshops and have grouped them under generic headings. This list is by no means exhaustive and, before an organisation commits to a partnering and integrated teamworking approach to construction, its senior managers should discuss, check and ensure that their value criteria are not only clearly specified but that they are set out, shared and understood by all members of their organisation.

Clients have been looking for:

- price – absolute cost certainty at board approval; build to cost and programme; whole life cost; minimum 15% off room cost within twelve months
- timeliness – completion ahead of programme; complete no later than programme date; development cycle (brief to occupation) reduced by 20%
- quality at handover – each scheme delivered on (specific date) with zero defects and standardised handover procedures and after sales care; defect free and meet response times; deliver the show flat on time; 100% tenant satisfaction at handover with no delays, defects or problems
- quality in use – contract warranties signed up one month before start on site; reduced tenant complaints.

Other client value criteria have included:

- drive sustainable solutions – reduce costs and waste; increase recycling; invest in local labour
- perfect communication; accountability; full tenant continual consultation; better response from tenants when the repair is reported
- fully integrated supply team enabling us to innovate and improve
- leveraging the partnering relationship to broaden our business opportunities
- a challenging, enjoyable and satisfying experience.

Suppliers (including consultants, constructors and specialists) have been seeking:

- cashflow – a good payment record; payment on time; good cashflow position
- profitability – lots of profit; fair and reasonable profit; reasonable profit with overheads secured; increased profit through continuous improvement (shared rewards); guaranteed minimum profit level with a mechanism to improve
- continued workload – continuity of future requirements; early commitment; repeat business; predictable workload through customer satisfaction; predictability of workload in all aspects
- quality – establish the right level of quality; zero defects
- specific programme criteria – workable programme; local work;

prestigious clients; public sector/private sector; work in specific periods of the year (to even out peaks and troughs)
- ❑ minimising risks
- ❑ safety – provide a safe working environment; the safe management of specialist contractors to ensure they know what they are doing to the standard required in the time available
- ❑ ensure information is issued correctly and on time and meets with all parties' approval.

Other supplier value criteria have included:

- ❑ better working relationships; open with each other; trust each other; trust in the client/familiarity
- ❑ pride in the job – enhance reputation (be part of success); enhance the brand; a building we can be proud of
- ❑ gather and pool our knowledge base; learn new skills; optimise the benefits of modular construction
- ❑ customer satisfaction; create a beautiful project loved by residents
- ❑ referrals for other business
- ❑ understand more about *best practice*
- ❑ closer look at environmental issues
- ❑ no stress; no disputes; non-confrontational work.

Developing and setting out clear sets of value criteria for each organisation will ensure that the integrated team has a clear understanding of better value as it applies to each organisation. In the first stage of the initial partnering workshop these organisational objectives will be aligned into a set of mutual objectives which will form the basis of the partnering charter for the project or partnering relationship.

3 Partnering Advisers and Facilitators

It is unlikely that any organisation will initiate or commit to cooperative working as a procurement approach to add value without some form of guidance and support, probably from outside the organisation. This principle will hold good whether the organisation calls this approach partnering, integrated teamworking, strategic alliancing or frameworking. This support will probably extend from the choice of procurement strategy, through selection to the first meeting of the team. Once the integrated team is established, there may be a further requirement for external assistance with drafting contracts and developing the relationships and skills of the team members through a programme of training and workshops.

Before the advent of partnering-specific contracts such as the Project Partnering Form of Contract PPC2000 (The Association of Consultant Architects Ltd & Trowers & Hamlins, 2000), the terms *partnering adviser* and *partnering facilitator* were often used interchangeably to describe consultants who offered support in partnering and integrated teamworking at various stages. The wide variety of best practice, process and people skills required in building the team, drafting contracts, resolving or mediating on issues and driving measured continuous improvement meant that the full service was rarely within the skill set of one individual.

The terms *adviser* and *facilitator* are now used separately to denote individuals (rather than organisations) who possess different skill sets and it is important for clients and constructors who are intending to follow the partnering and integrated teamworking route to first identify their specific support and consultancy requirements and then make the appropriate appointments.

It is critical to the success of the integrated team relationship that advisers and facilitators are not seen as favouring any organisation. Their fees and the costs of any workshop accommodation may be shared by the partners or included as a sum in the project budget.

The role of partnering adviser, as set out in PPC2000 and in the accompanying guide (Association of Consultant Architects Ltd & Trowers & Hamlins, 2003), does not feature in traditional contracts. Many of the tasks which the adviser carries out are new tasks arising as a direct consequence of the partnering approach – for example, the requirement to provide support in resolving issues. Other duties may have previously been carried out by the client representative, project manager or other project team member – for example, preparing contract documentation. It is important, therefore, that the core group ensures that the duties of the individual who carries out the role of partnering adviser (or whatever terminology is used in the specific contract) are very clearly set out in accordance with the appropriate contract.

It is important to ensure, before appointment, that the partnering adviser has practical knowledge of the specific contract being used as well as experience of the principles and practice of partnering and integrated teamworking. The Association of Consultant Architects (www.ACArchitects.co.uk) manages an Association of Partnering Advisers and maintains standards through examination of an individual's knowledge of PPC2000.

Whilst it would be inappropriate to concentrate solely on PPC2000 in this book, it is important to set out some of the duties of the partnering adviser, or the equivalent in other partnering contracts. These are clearly set out in PPC2000 in Clause 5.6 and include:

- preparation of all partnering documentation including review of specialist contracts to ensure consistency
- advice and support in relation to the operation of the contract
- support in resolution of an issue escalated above the highest level in the issue resolution ladder. As this may occur at any time it is necessary to ensure that the partnering adviser has adequate and appropriate back-up resources.

The role of the partnering facilitator is to facilitate meetings and workshops that the core group considers appropriate in the development of an integrated team and in support of project objectives. These may include teambuilding and joint training, initial partnering workshops, value and risk management, continuous improvement and post project reviews.

The role of the facilitator includes:

- impartiality (also removing the need for the project manager or another to take on the role)
- building the integrated team
- helping the team to come to decisions
- aligning the team's effort towards a common set of objectives
- developing a project culture from the separate cultures of the various organisations
- adapting workshop and facilitation styles to suit team dynamics
- valuing and encouraging the contribution of all team members
- maintaining the momentum of the workshop and delivering the stated objectives on time
- acting as a catalyst, challenging the team's thinking
- recording decisions and actions, writing and circulating clear reports promptly.

Note that the role of the facilitator is not to provide technical advice, to offer solutions or to make decisions on behalf of the integrated team.

In the early days of the relationship, the success of partnering and integrated teamworking may depend heavily on the skills and experience of the facilitator. As the team relationship develops and matures, processes are set up and the team develops a proactive approach to continuous improvement, the team's need for a facilitator may diminish. However, the partnering facilitator or adviser may still be called upon to assist with the resolution of issues at any stage throughout the relationship and to carry out reviews of progress. We have, for example, been asked to carry out annual reviews of term contracts with partnering teams although we do not have day-to-day contact through the rest of the year.

We have mentioned above that the partnering advisers and partnering facilitators should be independent of any of the partners and should be paid by the team. However, some client and contractor organisations have project managers with facilitation skills and may draft them in from other projects to facilitate workshops and help build the integrated team. The issue of corporate independence is slightly less critical in the case of the partnering facilitator than in the case of the partnering adviser, provided that all team members accept that the facilitator can, and does, demonstrate impartiality. However, carrying out the dual roles of project manager and facilitator may be difficult as there may be a perception of bias and there will be a need to change management styles from dynamic leader to coach and facilitator. The duality of roles could be confusing for team members.

The core group should consider the comparative values of in-house and external facilitators for partnering, value and risk workshops. There are advantages and disadvantages to both approaches and these should be carefully considered. We have listed below some of the advantages and disadvantages of using external over in-house facilitators:

Advantages

- independence and lack of bias
- less likely to be hindered by political considerations
- greater ability or freedom to challenge the status quo
- the ability of the core group to hire a specific skill at a specific time
- experience from other sectors/relationships

Disadvantages

- the fee may have be paid out of the project budget. Whilst the in-house facilitator's costs may not be charged to the project there will, nevertheless, be a resource cost which should be considered
- perceived lack of knowledge of specific organisations

The Integration Support Network (www.integrationsupportnetwork.org.uk) identifies potential facilitators by UK geographical region who offer a wide variety of facilitation styles and skills. The unregulated list was set up to provide training and consultancy support nationally in the UK resulting from the development of the integration toolkit mentioned in Chapter 1. The Institute of Value Management (www.ivm.org) can provide a list of qualified value management facilitators.

In our view, the two roles of partnering adviser and partnering facilitator are likely to be fulfilled by different individuals with different team role skills. The role of partnering adviser is suited to someone with a keen eye for detail and the detached overview to ensure consistency across all documentation. The role of partnering facilitator is suited to someone who can challenge and drive the team forward and is extrovert and keen to explore new possibilities.

4 Internal Partnering and Managing Change

The most significant barrier to the successful development of a partnering and integrated teamworking approach in the delivery of better value is the cultural barrier – the various traditions and processes within organisations and the attitudes of individuals that have all become established over time, with or without clear rationale or business case.

Before any organisation can consider itself ready to partner with another, it must ensure that it is partnering internally, that individuals and departments are working for the common good of the organisation. In particular, the organisation's management team should address:

1. whether corporate objectives have been clearly established and communicated (if they have not defined where they want to be, they cannot partner to achieve it)
2. the degree of understanding and buy-in to corporate objectives
3. the degree of alignment of departmental objectives with corporate objectives
4. a lack of cooperative working up and down the management structure (evidenced by a lack of trust and empowerment) and
5. a lack of cooperation across departments, each of which may be tasked with targets that are not aligned with the targets of other departments.

Organisations that depend on a rigid hierarchical management structure, operating a *tell* culture rather than an empowered workforce, will find it more difficult to implement a successful partnering or integrated teamworking approach. For example, if a senior manager insists on making all the decisions on a project, the project manager may simply be acting as a post box. This will lead to delay in resolving issues as it will be necessary to allow the issue to pass through the project manager's hands to the senior manager and then back, via the project manager, down the chain of command. The organisation will also be faced with the wasted cost of both a project manager and a senior manager being involved in the resolution of the same issue.

Any organisation that intends to drive better value by changing to a cooperative approach to the procurement of construction must also ensure that the departments within its own organisation understand each others' key objectives. Management must align these objectives and encourage the departments to work cooperatively, changing departmental processes, attitudes and traditions for the common good. For example, it is of little benefit if a client's development department agrees 14-day payments with a constructor if the finance department sends out cheques by second class post, 28 days after receipt of an invoice. Both departments may have their own objectives (development may be seeking lower capital costs and finance may be seeking to earn interest from the money on deposit) but these objectives may conflict. The internal silo culture must be bridged before the external approach to integrated teamworking can be successfully implemented (see Fig. 4.1).

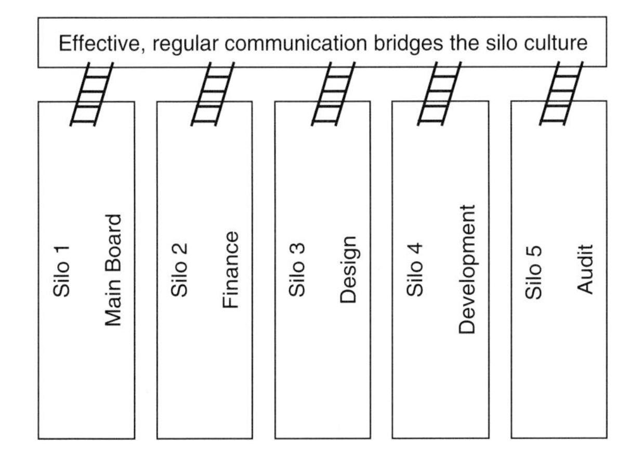

Figure 4.1 The silo culture.

The first stage in bridging the silo culture must be for each department to understand the value criteria of the other departments and identify where specific objectives are in conflict. This is best carried out in a cross-departmental partnering workshop at which senior members (decision makers) from all departments should be present. We would base a typical agenda for such a workshop on the agenda for an initial partnering workshop, set out later in this book. In outline, the stages to be covered are:

- introductions and objectives of the day
- commitment to cross-departmental partnering by board member
- develop mutual objectives, aligning departmental and corporate objectives
- agree issue resolution processes within and across departments
- identify areas for continuous improvement after an informal gap analysis comparing *where are we now?* with *where do we want to be?*
- agree key performance indicators based on the mutual objectives
- agree implementation plan.

The output from the workshop should be cascaded by those present to the remainder of the organisation, along with the agreed actions to develop the culture of cooperation. Clear leadership, good communication and ownership of the change process will smooth the transition from a silo culture to an internal partnering culture. This will give the organisation a better chance of developing successful partnering and integrated teamworking relationships with others.

Changing the corporate strategy from lowest price procurement to best value may be considered as a revolution in an organisation's culture. The fact that an organisation has previously been successful in adopting a lowest price culture may make it difficult for directors, management and staff within the organisation to accept the change to a best value approach ('if it ain't broke, don't fix it'). It is necessary for senior management to identify and communicate the reasons why the change is necessary and to drive the change process, encouraging all members of the organisation to buy into the new ways of working. It may be important to make clear the reasons why the strategic objectives have changed – for example, whether this is as a result of external pressures (regulation or a changing industry) or internal priorities (greater productivity and added value).

Those at the top of the organisation must own the drive for change. If the main board does not see an advantage in a change to a value-based, partnering and integrated teamworking approach, then the initiative will not succeed. The advantages and disadvantages, benefits and costs must be set out clearly so that the board or senior management can not only understand the rationale for the change – the business case – but can also communicate this through the organisation. The most successful partnering arrangements that we have facilitated have been those in which senior management have established and maintained a structured programme to communicate the vision, structure and objectives of the relationship supported by board members and other interested parties such as auditors and elected members.

Ironically, change is a constant in most organisations in order to maintain competitive advantage. Many individuals within these organisations will feel that they have had little time to implement one initiative before another comes along. Management need to be sympathetic to the potential for initiative overload and consider whether they can remove one task before adding another. For example, before moving to a best value selection process, management should remove the standing order that insists on appointing suppliers on the basis of lowest price competitive tender.

We have identified six stages to the management of effective change. The six stages are summarised in the acronym CHANGE:

- Challenge – initiate a challenge to the status quo, identify the criteria (internally led or externally applied) which may be forcing the change and involve interested parties.
- Hypothesise – treat change as a project, select an appropriate change team, develop the thoughts and the business case in joint departmental workshops, identify the positive and negative potential impact on the organisation and on individuals, agree the objectives and the programme and appoint change champions for each department.
- Act – share the objectives and programme, break down barriers and involve the wider cross-departmental team in implementing the change, train management and staff, listen to and address concerns and fears and ensure the change champions meet regularly.

- ❑ Nurture – encourage individuals, avoid overload by removing one task or process for each new task added, log and publicise benefits to bring on board any cynics.
- ❑ Grow – persist as culture change takes many years to bed in and become the norm, maintain the training regime and induct new members into the team with a full training package.
- ❑ Evaluate – compare results with predictions, review and refine processes and refocus the team on a regular basis.

We are fortunate to have assisted United House Ltd (UHL), a social housing contractor, in the early stages of managing change in the organisation, taking up the challenges posed by the, then, recently published, *Rethinking construction* (Egan, 1998). UHL's change was owned and driven by Jeffrey Adams, Managing Director of the company, who initially set out a clear vision to the directors. We worked with Paul Greenwood (Best Practice Director) to develop a visioning and communications programme to ensure that all departments understood corporate objectives and the principles and practices of partnering as set out in *Rethinking construction.*

In the early days of the process, we facilitated a workshop at which directors of all departments clarified corporate goals, developed an understanding of how partnering would assist in achieving these, defined key performance indicators (KPIs), agreed behavioural attributes of a partnering culture, identified partnering champions and defined a process for communication to the management and staff of the organisation.

Two workshops followed in the next month to involve senior management by sharing the partnering culture, communicating partnering goals and KPIs, agreeing practical steps for senior management and staff to implement KPIs, identifying and resolving issues and proposing and developing project partnering processes at all stages from brief to final account.

Over the next three months these workshops were followed by three training workshops for all site management and staff. These ensured common understanding of the principles of *Rethinking construction* and the United House Charter by sharing the partnering culture, communicating partnering goals and KPIs and sharing forthcoming changes in the industry (partnering and Egan-compliance). The delegates were also tasked with agreeing practical steps for site management and staff to implement partnering and KPIs, to identify and resolve issues, and to propose and develop project partnering processes at all stages from brief to final account. Members of UHL who were unable to attend the formal training workshops received detailed briefings from their directors and Paul Greenwood. Although partnering is now embedded in the business culture, all new staff, whether labourers or directors, receive an introduction to *Rethinking construction* as part of their company induction – there is no escape!

A year after the initial directors' workshop we facilitated a workshop exploring partnering and cooperative working for subcontractors and management of UHL and the programme was completed with two further workshops for directors to review the earlier work, assess the impact and develop further strategies.

Following the initial training, the company has further developed its processes on customer focus, recruitment, induction and many other aspects of *Rethinking construction*. These led to United House Ltd gaining the Building Homes Quality Award 2003 for Best Change Strategy and have enabled them to maintain a regular flow of quality work from clients who can identify the value of partnering and integrated teamworking. With over 80% of work now based on partnered contracts of all flavours, the company is signing long-term strategic relationships with equally innovative local authorities and registered social landlords.

The added value that an organisation derives from partnering and integrated teamworking depends on the extent to which it can apply and develop the principles of partnering and integrated teamworking within and across its own departmental structure. All members of the organisation, from labourer to director and head office to site, must demonstrate a value-based, cooperative culture if the organisation expects to develop a similar culture within the wider partnering and integrated team.

5 Selection Criteria and Weighting

In *Constructing the team* (Latham, 1994), Sir Michael Latham suggested in paragraph 6.39 that, '... those tenders which offer the best value for money ("economically advantageous" in EU terminology) and show clear regard for cost-in-use should be accepted.' He also identified that, '... public authorities should publish their own criteria for quality assessment in their tender documents ...'.

To achieve better value, all clients, whether in the public or private sector, should choose their constructor with appropriate weightings of selection criteria, rather than automatically accepting the lowest tender. Lowest tendered price does not necessarily mean best value as there may be occasions when a supplier (whether this be consultant, constructor or specialist) chooses to price low to win work, regardless of their profit. They may then concentrate on trying to regain lost margin through use of cheaper materials and/or seeking and exploiting loopholes in the contract documentation, rather than concentrating on carrying out the work as specified. Thus the client and their representatives may not get what they specified and may also have to expend valuable resource on one-on-one marking, policing the work off and on site and fighting their corner.

A structured selection process, based on clear value criteria and a robust scoring method for qualitative and quantitative criteria, is the foundation of selecting an effective integrated team, fully aligned and focused on delivering best value for all concerned. This process must be developed and implemented at the very start of the decision to appoint the team. However, whilst this process may be perceived as a client's process to select suppliers, it must not be forgotten that suppliers also have a choice and it is likely that they will evaluate the clients that they want to work with on the basis of a similar structured process. We have therefore set out separately the two processes below, using the generic term 'suppliers' to refer to constructors, consultants and specialists.

CLIENTS SELECTING SUPPLIERS

The client's supplier selection team should set aside an appropriate time to review the client's value criteria and develop the selection process and scoring matrix. In our experience, this is unlikely to take less than a day as there will be considerable discussion on the relative weightings of each criterion and the team should be allowed sufficient time to reach consensus.

In order for a client's supplier selection team to make a robust value decision on the selection of a supplier, they must set out their key value criteria, prioritise and weight these according to their relative importance. Once the weightings are agreed, the team can develop questions that will enable them, transparently and objectively, to score the submitted responses of each supplier in each criterion. The criteria weightings and supplier scores will be placed on a matrix which will extend and total the scores of each supplier.

In order to ensure the appropriate weightings, the client's supplier selection team must first agree the appropriate selection criteria. It is very likely that these will be based on the client organisation's value criteria. The team will develop the selection criteria and we would emphasise the importance of cross-departmental representation and input from other interested parties such as:

- end-user departments (e.g. housing management, library management, hotel managers)
- maintenance
- residents
- audit and/or best value inspectors
- accounting and finance.

The facilitator should first elicit the appropriate value criteria from the members of the client's supplier selection team (e.g. price, quality, speed, cooperation, etc.). At this stage, the team should list all criteria that they feel are appropriate – the reduction to a more manageable number can come later. Clarity of definition is critical to the later development of a scoring process. Therefore the team should ensure that each criterion can be clearly and objectively judged. Thus *time* will not be sufficiently clear but *shortest build period* or *shortest time from brief to handover* may be a more appropriate term depending on the client's view.

Having created a list of all value criteria, the client's supplier selection team should consider the number of selection criteria that will be appropriate in the selection process. We suggest no more than ten. The team should bear in mind that the average percentage weighting for each of ten criteria will be 10%. As price may account for 40% or more of the total, the remaining nine quality criteria will account for 60% = 6.7% average for each one. Any weighting of less than 4% is unlikely to have an impact on the final order of assessed bids and may be discounted or rolled up into another criterion.

The facilitator should distribute a copy of the shortlist of the agreed selection criteria to each person, asking them to apportion 100 points across the ten categories. At this stage, the client's supplier selection team members are working in deliberate isolation and the facilitator should allow adequate time for each person to allocate, review, revise and finalise their weightings. At the end of the individual session, the facilitator should set up a flip chart or (preferably) a spreadsheet with data projector and ask each team member to give their weightings for each criterion (thus all weightings for criterion one will be collected at the same time). There may be variety in the weightings but discussion should be suppressed until all are entered. At this stage the team will be ready for a break as, in our experience, it may take up to two hours to complete the shortlist of selection criteria and set out individual weightings. After a break, the facilitator will lead full team discussion and negotiation between team members to arrive at consensus on a single weighting for each criterion.

The process so far has identified:

- ❏ the key value criteria that will enable the client's supplier selection team to appoint on best value
- ❏ the weighting of each criterion as a percentage of the total.

SUPPLIERS EVALUATING CLIENTS

In order for suppliers to make a robust value decision on which client(s) and client representative(s) they will choose to work with, they must set out their key value criteria, prioritise and weight these according to their relative importance. Once this is agreed, the suppliers can develop questions that will enable them objectively to evaluate their perception of that client within each criterion. The criteria weightings and client scores will be placed on a matrix which will extend and total the scores for each client.

In order to ensure the appropriate weightings, the supplier must first agree the appropriate evaluation criteria. It is very likely that these will be based on the supplier organisation's value criteria. Evaluation criteria will be developed by the supplier's client evaluation team and it is important that input is sought from interested parties such as:

- estimators
- domestic subcontractors
- site management
- surveyors
- after-sales or defects management team
- accounting and finance.

The supplier's client evaluation team should set aside an appropriate time to review their value criteria and develop the evaluation process and scoring matrix. In our experience, this is unlikely to take less than a day as there will be considerable discussion on the relative weightings of each criterion and the team should be allowed sufficient time to reach consensus.

The facilitator should first elicit the appropriate value criteria from the supplier's client evaluation team (e.g. profitability, location, information flow, cooperation, etc.). At this stage, the team should list all criteria that they feel are appropriate – the reduction to a more manageable number can come later. Clarity of definition is critical to the later development of a scoring process. Therefore, the team should ensure that each criterion can be clearly and objectively judged. Thus *location* will not be sufficiently clear but *within an hour's drive* or *within the M25* may be a more appropriate term, depending on the supplier's view.

Having created a list of all value criteria, the supplier's client evaluation team should consider the number of evaluation criteria that will be appropriate. We suggest no more than ten. The team should bear in mind that the average percentage weighting for each of ten criteria will be 10%. Any weighting of less than 4% is unlikely to have an impact on the final order of client preference and may be discounted or rolled up into another criterion. The facilitator should distribute a copy of the shortlist of the agreed evaluation criteria to each member of the team individually to apportion 100 points across the categories. At this stage, the team members are working in deliberate isolation and the facilitator should allow adequate time for each person to allocate, review, revise and finalise their weightings.

At the end of the individual session, the facilitator should set up a flip chart or (preferably) a spreadsheet with data projector and ask each supplier's client evaluation team member to give their weightings against each criterion (thus all weightings for criterion one will be collected at the same time). There may be variety in the weightings but discussion should be suppressed until all are entered. At this stage, the team will be ready for a break as, in our experience, it may take up to two hours to complete the shortlist of evaluation criteria and set out individual weightings. After a break, the facilitator will lead full team discussion and negotiation between team members to arrive at consensus on a single weighting for each criterion.

The process has so far identified:

- ❑ the key value criteria that will enable the supplier's client evaluation team to evaluate clients and client representatives on best value
- ❑ the weighting of each criterion as a percentage of the total.

The following example is based on a number of supplier partner selection processes that we have facilitated. Whilst it is a summary of a complete process developing weighted criteria, selection teams should ensure that they use this only as a guide, using their own value criteria and following appropriate sector-specific and organisational procedures and processes.

We were appointed to facilitate a meeting of the client's supplier selection team in order to set weighted objective criteria to select constructor partners. The selection team was asked to set aside the whole day and informed that lunch would be provided.

After discussion on the client's value criteria, as set out in their mission statement, we asked the selection team members to identify the criteria that they would use in selecting constructor partners. A list of ten criteria was agreed. Note that this particular selection process did not include price. These quality criteria were to be used to provide 50% of the available points and the price analysis would provide the remainder.

We printed out the list of criteria, one copy for each selection team member, and asked each person to allocate 100 points across the 10 criteria. This stage took 40 minutes. All team members were encouraged to double check their prioritisation by, for example, asking them if they thought that safety was more or less important than zero defects. In this way, a considered view was elicited from each selection team member.

In the feedback session, it was identified that different team members had different priorities and applied different weightings to each criterion. It was therefore necessary to discuss the reasons and to reach a consensus. This task took the remainder of the day. The agreed weightings were as shown in Fig. 5.1

Knowledge of sector	5, 15, 8, 10, 8	=	**9%**
Partnering credentials	20, 15, 13, 15, 15	=	**16%**
Safety and KPIs	10, 5, 15, 20, 5	=	**11%**
Environment and sustainability	5, 5, 9, 5, 12	=	**7%**
Value management	5, 10, 5, 10, 10	=	**8%**
Commitment to budgets	15, 10, 14, 15, 15	=	**14%**
Quality	15, 15, 14, 5, 12	=	**12%**
Supply chain management	5, 10, 4, 3, 9	=	**6%**
Off site manufacture	10, 10, 4, 2, 9	=	**7%**
Company profile	10, 5, 14, 15, 6	=	**10%**

Figure 5.1 Partner selection criteria weightings.

Before any organisation (client or supplier) can deliver best value from a partnering or integrated teamworking relationship, it is critical that they clarify their own value criteria so they can select appropriate partners who have a proven track record in their key criteria. A structured evaluation and selection process, based on scores against weighted value criteria, is the foundation for building an effective integrated team, fully aligned and focused on delivering best value for all partners.

6 Selecting Supplier Partners

After identifying the client organisation's value criteria, the client's supplier selection team have prioritised and weighted the shortlist of selection criteria. These weightings are now set and will not be changed during the selection process. We make the point about not changing the weightings for two reasons. The first is to impress on the team the importance of agreeing the weightings at an early stage and to ensure that appropriate attention and resource is given to the process. Secondly, we make the point very forcefully to discourage teams who, at bid evaluation, might be tempted to change the percentages around a bit to get the supplier they thought they wanted. Ensuring the proper focus at the early stages will make bid assessment, evaluation and award much more straightforward and productive and ensure that selection teams stay within the bounds of good practice and the law.

The client's supplier selection team is now at a critical stage in developing a successful structured value-based selection process. This stage involves setting the programme for the selection process, agreeing and developing a scoring system for supplier responses and finalising questions to include in the partnering documentation. The questions will need to draw out appropriate information from the supplier to enable the team to evaluate and score each supplier on their performance in each of the selection criteria. We address the programme and scoring system in this chapter and set out the approach to eliciting supplier information in the next chapter.

The client's supplier selection team should bear in mind that our suggestions above do not take account of any specific organisational or sector rules which the team must follow. However, it is always worth the team taking time to check and ensure that the current interpretation of the rules is correct. We have found that standing orders may be cited as a reason for using price-only tenders but standing orders should, by now, have been changed to *most economically advantageous* (e.g. including whole life costs and quality aspects) so it is worth checking the small print.

The structure of the programme for supplier selection will depend first on the number of suppliers the client team wishes to appoint at the end of the process. For single capital projects or term contracts a single appointment may be made. For framework contracts, where a programme of work is shared between a number of suppliers, five or six suppliers may be appointed. In either case, we prefer to set up programmes working back from the number of supplier partners finally required.

For the appointment of a single supplier, the next stage is to determine how many suppliers can be realistically compared at the shortlist stage. Under traditional tendering guidelines, six suppliers would probably have tendered a bid. The amount of work that a supplier is required to carry out in order to win a partnering contract is higher than that required in a price-only bid so we would suggest a smaller number, probably between three and six. We understand that some organisations feel that three is too small a number to bid for purposes of probity but we disagree with the viewpoint that greater numbers means greater probity. Provided the shortlist has been put together through a structured and transparent process based on clear value criteria, there should be no reason to increase the number of tenderers just to reach a predetermined number. An increased number of tenderers results in increased waste in the costs of preparing documentation and tendering for both the client and the suppliers. There must be trust in a partnering relationship and the selection process is a good place to start to demonstrate this trust and to reduce the wasted costs of bidding. For frameworks, we suggest that the number of suppliers at the shortlist stage could be a multiple of two to three times the number of eventual suppliers as shown in Fig. 6.1.

In order to arrive at a shortlist, the client's supplier selection team will need to create a longlist of interested suppliers. The number of suppliers on the longlist will depend on whether the bid process is to be open to all interested parties (as in some public contracts) or limited to those suppliers already known to the client organisations (as in some private contracts). In some cases the longlist may be

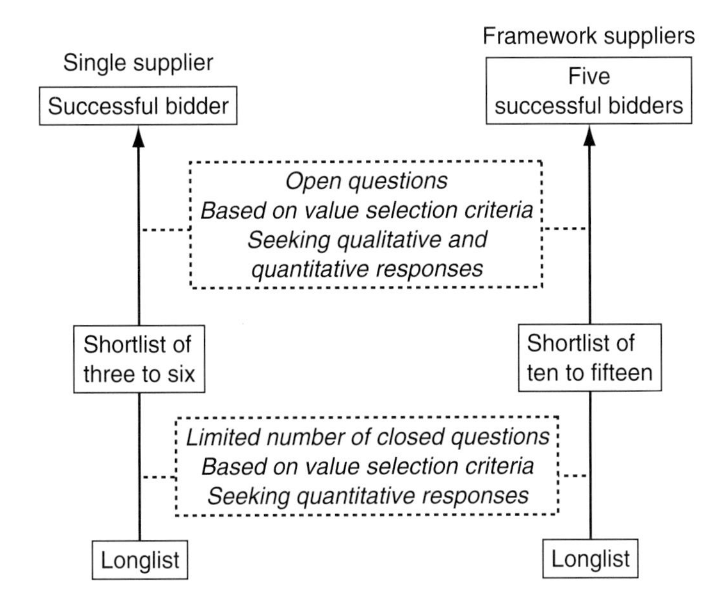

Figure 6.1 Supplier selection process.

dispensed with – for example, if the number of interested suppliers is already below the appropriate shortlist number.

Once the client's supplier selection team has drawn up the number of suppliers at each stage (for example, three partners, eight on the shortlist, fifteen on the longlist) the team should set out the programme for reducing the longlist through the shortlist to the final number. If sufficient time is to be given to suppliers to prepare reasoned responses this is unlikely to be a short process and the team should give robust consideration to the programme. We have outlined below a programme assuming a selection process for a framework of three suppliers with a shortlist of eight and a longlist of fifteen.

LONGLIST OF POTENTIAL PARTNERS

Initially, the client's supplier selection team must create the longlist. For organisations that must advertise their projects (through European, sector or organisation-specific rules) the creation of this longlist is out of the hands of the team. For other organisations, the list may be drawn up by previous experience, local knowledge, a desire to break the mould or any other rationale. However many names are on the longlist, the first stage selection process will identify and shortlist only the top performing suppliers against the client's selection criteria.

In the process from longlist to shortlist, the client's supplier selection team should first check that the organisations are willing to have their names put forward, as some may not want to work for that specific client. The team should then prepare a limited questionnaire of mainly closed questions that will elicit objective data for comparison. At this stage, after due consideration of probity, ease of comparison and speed of selection are the priorities. The team should set questions based on the selection criteria identified earlier and develop a matrix on which to enter each supplier's scores.

At the longlist stage, the client's supplier selection team may consider offering short interviews to each of the potential partners. In one such process, we spent seven days at a central venue interviewing twenty-eight longlist constructor organisations for one hour each at two hour intervals. We had prepared a set of questions to cover the first thirty minutes and then opened the session to the constructors to raise their own questions. Each constructor was scored at the interview on how they performed against set criteria. The team specifically asked for tradespeople to be present rather than marketing executives, assessing how all constructor representatives might work in an integrated team environment and how the ethos of partnering had been cascaded throughout the organisation.

On the date for return of the longlist questionnaire, the client's supplier selection team should meet to assess and score the responses. The scoring process should be set up to ensure an objective quantitative assessment of suppliers. We cover the process of question setting in detail in Chapter 7 and assessment in Chapter 10. The top scoring suppliers will be offered the opportunity to go forward to the shortlist and a representative of the team will write to the unselected suppliers, informing them of the situation and offering to share with them the results of the longlist process.

SHORTLIST OF POTENTIAL PARTNERS

In the process from shortlist to selected partner(s) the timeframe will be more extended. The client's supplier selection team will develop a more extensive questionnaire based on the previously identified selection criteria and may want to set up a more comprehensive face-to-face meeting with more members (management and staff) of the supplier organisation.

At the shortlist stage, the client's supplier selection team should consider visiting the suppliers at their offices and on site. Depending on the geographical spread of suppliers, this may mean a more extended submission period. The selection team should assess how much resource is available and what added value this brings before committing to site visits but we have found them invaluable on numerous occasions. Teams may choose to score head office and site visits as part of their selection criteria. A pre-agreed set of questions ensures that supplier responses can be compared. For one regional framework selection, the team of client representative, partnering facilitator and three client managers visited the head office and one site for each of two potential constructors each day across a period of four days. This proved to be a very tight programme and keeping to the programme required great team discipline.

Within the letter accompanying the final questionnaire/partner selection documentation, which may include pricing documents, the client's supplier selection team should set out the timeframe for return of the completed bid documents and how the client's team will respond to supplier questions before return of bids. The team should identify a single point of contact to ensure that all prospective partners are treated fairly and consistently.

Questions at the shortlist stage should be more open and seek a greater depth and breadth of information than at longlist stage. Accordingly, the time given to the potential partners for compilation of the bid should be extended. We believe that 50% more time is required for a partnering submission than would be usual under a price-only tender.

SCORING SYSTEM

In setting the scoring system for responses, some client's supplier selection teams and their partnering consultants use a percentage-based system. We have found this to be excessively complex. We find it difficult to ascertain objectively, for example, whether the score for a particular qualitative response is worthy of 63% or 67%. Our system (whether at longlist or shortlist stage) is based on a four-step evaluation of whether or not the assessors judge that the supplier's response meets the client's requirements.

0 = unacceptable
1 = acceptable with improvement
2 = acceptable
3 = exceeding expectations

This scoring system ensures that any supplier response that meets acceptable levels of performance scores two points in that selection criterion. A response that demonstrates unacceptable levels of performance scores zero points.

There will be some responses that may not achieve the acceptable (two points) standard, yet are not totally unacceptable (zero points). In such cases, one point may be awarded on the condition that the supplier takes effective steps to improve their performance within a reasonable period.

On occasion, the team may identify a supplier whose response is not only acceptable but is outstanding in one criterion, eliciting such expressions as 'world class', 'exceeding expectations' or 'wow!'. This supplier response will attract a score of three points. We would suggest that an assessor should credit only one supplier with a three-point score in any one criterion.

Some criteria will require the client's supplier selection team to decide whether a zero score will result in the supplier's bid being totally rejected – even if they were to score top marks in all other criteria. Such 'drop dead' criteria should be identified before bid documents are sent out and not left to a decision after the return of bids. Examples of 'drop dead' criteria might include a very poor record on health and safety or a low financial standing. For 'drop dead' criteria it should be incumbent upon the team to make initial enquiries of all prospective bidders to ascertain their record in these criteria. The early enquiry may save effort and resource on behalf of the bidder and the team.

The client's supplier selection team now has:

- ❑ a set of the client's value criteria
- ❑ a schedule of selection criteria based on the client's values
- ❑ weightings against each of the selection criteria

- ❑ a process to select partners from a longlist through to a shortlist
- ❑ a mechanism to compare and score suppliers' responses at longlist and shortlist stages.

During this stage of the process, the client's supplier selection team will have constructed a matrix enabling them to make an objective assessment on how well specific suppliers meet the value requirements of the client. Using this matrix will result in a single score for each supplier, identifying the best value supplier.

In Fig. 6.2 we have taken the matrix developed in the previous chapter, including the percentage weightings against each criterion, and scored a notional supplier in the 0–3 scale against each criterion. This has been extended and totalled, resulting in a single value score for that supplier.

Having set weightings and agreed scoring principles for the value based selection criteria, it only remains for the client's supplier selection team to agree and set questions in the partnering documentation for each of the criteria at each stage (longlist and shortlist) and to identify responses that will be unacceptable, acceptable with improvement, acceptable or exceeding expectations.

		Score		**Weighted score**
Knowledge of sector	9%	x 3	=	27
Partnering credentials	16%	x 2	=	32
Safety and KPIs	11%	x 2	=	22
Environment and sustainability	7%	x 1	=	7
Value management	8%	x 3	=	24
Commitment to budgets	14%	x 3	=	42
Quality	12%	x 2	=	24
Supply chain management	6%	x 1	=	6
Off site manufacture	7%	x 1	=	7
Company profile	10%	x 2	=	20
	Total score		**=**	**211**
	out of a possible			**300**

Figure 6.2 Partner selection criteria scores.

7 Writing Effective Partnering Documentation

Having identified the client's value criteria and set weighted supplier selection criteria, the client's supplier selection team must now set questions that will assess suppliers' performance against the selection criteria. The setting of the questions may be best undertaken by individual members of the team depending on their technical skills. This will be followed by a team review, discussion and agreement. For each of the selection criteria the team should set out clear measures that will provide an auditor with the proof that they have clearly defined the difference between unacceptable and acceptable performance. These measures may also form the basis of determining, during the contract, whether the performance of the integrated team is acceptable.

In the event that the integrated team's performance falls below acceptable standards during the contract, the core group will seek to apply corrective measures or, in extreme circumstances, they may choose to determine the contract or partnering arrangement. The client's supplier selection team should, therefore, at an early stage in developing their partnering and integrated teamworking strategy, formulate an exit strategy for any partner or member of the integrated team. This is a particular issue for public sector organisations whose arrangements are limited to a fixed period by legislation.

The exit strategy should be set out clearly in the selection documents and agreed with prospective partners to ensure that all partners agree both the specific event(s) that will lead to a determination of the arrangement and the processes that will be followed by all partners in ensuring a determination without rancour. All partners should ensure that the exit strategy preserves the spirit of trust that marked the start of the relationship and ensures the ongoing maintenance of commercial confidentiality and intellectual property rights. We have identified three triggers for the determination of a partnering or integrated teamworking relationship. Processes should be considered and agreed for each one before the arrangement is set up.

1. expiry of the statutory period
2. consistent performance below agreed and clearly specified levels by one or other of the partners
3. an agreement by both partners that the arrangement must be determined, for example, triggered by external factors outside the control of the partners.

In moving from the longlist to the shortlist of potential partners, it is important that the client's supplier selection team sets a longlist questionnaire that is in line with the original value criteria of the client and the agreed selection criteria. At this stage, the questionnaire should comprise closed questions seeking concise responses. Other requests for information can be made in the form of statements but should still seek concise responses that can be quickly and easily assessed.

The following example is included as a basis for client's supplier selection teams in developing their own longlist partnering selection questionnaire. In addition to setting the questions, the team should inform the potential suppliers of the scale of the project/programme, the selection criteria and the weightings for each criterion. They should also inform the suppliers of the way in which the scoring will be carried out (in our example we have used the 0–3 scale). We have set out typical wording for this in the shortlist questionnaire example at the end of this chapter. Note that we have included typical questions against only three of the ten criteria that we identified in earlier chapters.

Longlist questionnaire

1. Partnering credentials

   Do you have experience of partnering in this (e.g. social housing, hotel) sector? Yes/No

   Do you have experience of partnering in any other sector? Yes/No

If the answer to either of the above questions is positive, please provide names of the client and lead consultant, the contract sum, site duration in weeks, date of handover, final account and the name of your contracts manager for two recent partnering contracts.

2. Value management
   Identify two recent value management/engineering proposals that you have developed and quantify the added value for each.
3. Off site manufacture (OSM)
   Do you have experience of OSM in this (e.g. social housing, hotel) sector? Yes/No
   Do you have experience of OSM in any other sector? Yes/No
   If the answer to either of the above questions is positive, please provide details of the client, lead consultant and OSM specialists on two recent OSM projects.

Having set the questions, the client's supplier selection team should identify the responses which will generate a specific score in the 0–3 scoring system that we have outlined. These should be set out on a single sheet that can be handed out to the assessors. Examples for guidance are set out below.

Response scoring sheet

Partnering credentials

0 No experience of partnering in any sector
1 Experience of partnering in any sector
2 Experience of partnering in this sector or
Experience of partnering in any sector and requested information on two projects
3 Experience of partnering in this sector and requested information on two projects

Value management

0 No experience of value management
1 Experience and case study but no quantification of added value
2 Experience and quantification of added value in two recent projects, both of which are cost reduction
3 Experience and quantification of added value in two recent projects, one of which shows added value in ways other than cost reduction

Off site manufacture (OSM)

0 No experience of OSM
1 Experience of OSM in any sector
2 Experience of OSM in this sector or
Experience of OSM in any sector and requested information on two recent projects
3 Experience of OSM in this sector and requested information on two recent projects

In moving from the shortlist of potential partners to the final selection, it is important that the client's supplier selection team sets a shortlist questionnaire that is in line with the original value criteria of the client, the agreed selection criteria and the earlier longlist questionnaire.

Questions should be set in such a way that suppliers have to use and demonstrate the full depth and breadth of their knowledge and expertise in responding. Closed questions that can be answered with a simple positive or negative response may, at this stage, be of limited use. Open questions are preferable. For example, on the topic of key performance indicators (KPIs), the question, 'Do you prepare radar charts of KPIs on all your projects?' is a closed question that can only attract a positive or negative response from suppliers. This question does not seek specific deliverables or enable the selection team to make any but the most basic differentiation between suppliers. The open question/statement, 'Supply the KPI radar chart for your most recent project with calculations, analysis and observations for each indicator' will identify those who are familiar with the process of preparing KPI charts and ensure that the selection team has a project-specific radar chart to enable them to analyse actual performance in more detail.

The questions should be set in such a way as to enable the client's supplier selection team to cut through marketing hype. The team should bear in mind that, in responding to selection processes, suppliers may pass the qualitative submission documentation to a marketing department whilst the estimators get on with the job of pricing the quantitative documents. Questions should be set so that suppliers have to carefully consider their responses and the questions should attempt to get behind the information presented in standard marketing pages. For example, we would question the value of asking for

1. Partnering credentials

The client will not tender future partnered projects. Budgets will be developed and agreed with the selected partner. Bearing in mind that this is a major demonstration of trust...

How will you demonstrate and ensure high levels of predictability and budget certainty?

What will you require from the client and consultants in order to deliver this certainty?

2. Value management

Considerable effort is being made by the client's design team to refine and improve the product to drive further customer satisfaction, reduce defects and waste, reduce cost and time on site.

How would you structure and/or contribute to this programme of continuous improvement?

3. Off site manufacture

You will be aware of the need for the client to incorporate off site manufacture into its projects.

What steps would you take to ensure that your projects are constructed to the tolerances required of a manufactured solution?

Having set the questions, the selection team should identify the responses which will generate a specific score in the 0–3 scoring system that we have outlined above. These should be set out on a single sheet that can be handed to the assessors. Examples for guidance are set out below.

Partnering credentials

0 All responses focused on the client and consultants delivering the right information at the right time

1 Proposals for budget certainty supported by evidence of previous integrated teamworking to achieve goals

2 Proposals for budget certainty supported by evidence from a specific project with a continuous improvement process applied with full integrated team attendance

3 Proposals for budget certainty supported by quantified evidence, on a named project, of specific improvements arising from a continuous improvement process with full integrated team attendance, enhancing value for all partners.

Value management
- 0 No proposals 'We'll build what you tell us' or 'This is our product...'
- 1 Some limited proposals mainly focused on improved client or specialist performance
- 2 Proposals focused on team performance. Clear understanding of structured VM process
- 3 Unsolicited recommendations made to improve the client's product or process

Off site manufacture (OSM)
- 0 No awareness of OSM or misunderstanding of manufacturing pressures
- 1 Awareness of the difficulties – 'we'll cross this bridge when we come to it'
- 2 Clear proposals for integrating prefabrication into the design and construction process
- 3 Already prefabricating successfully with quantified proof of benefit

The client's supplier selection team have now set questions that will assess suppliers' performance against the weighted selection criteria at longlist and shortlist stages. For each of the selection criteria, the team has set out clear measures that will enable assessors to score and rank suppliers whilst providing auditors with evidence that they have clearly defined the difference between unacceptable and acceptable performance.

8 Evaluating Client Partners

In contemporary value-based selection processes, clients are providing constructors, consultants and specialists (for whom we will use the generic term suppliers) with an opportunity to bid for work not solely on the basis of lowest price. Those clients in search of better value will produce tendering documents that allow suppliers to differentiate themselves from their rivals on clear quality criteria and provide them with the opportunity to propose alternative added value solutions within their bids.

Partnering and integrated teamworking produces the greatest added value to all parties, including suppliers, when the contract value is sufficient to warrant the additional early resource input (for example, to partnering, value and risk management workshops) and when the team works effectively and efficiently together.

Suppliers will, therefore, adopt a variety of responses to potential clients depending on the suppliers' evaluation of the client. The responses will depend not only on whether the contracts are sufficiently large and profitable, but also on whether the cultures of the client and others involved make for a rewarding contracting experience.

Any supplier organisation can win work on a price-only contract if they are prepared to buy the work and the quality of submission does not come into the tender equation. The supplier won't necessarily make any money on the deal but the contract is there for the taking. However, having won a loss-making contract, the supplier may concentrate on subsequent opportunities to regain lost margin by identifying, for example, loopholes in the tendering documents. This approach is heavy on the requirement for the supplier to provide post-contract resource such as claims surveyors and solicitors. This tends to drive one-on-one marking from the client, increasing the resource expenditure on the contract and possibly leading to a greater expenditure on resources than the perceived 'saving' in the contract sum. Nobody wins consistently in this approach.

Winning quality and price, value-based contracts requires skill, judgment and an ability to communicate the skills and unique attributes of the supplier's organisation – the added value items that make the organisation stand out from those who will also be bidding. The higher the quality proportion of the scoring matrix, the more this will apply. Winning a profitable contract with an appropriate (not necessarily lowest) price by outperforming rivals in the quality aspect of the bid, will reduce the requirement for the post-contract resource, previously used to regain margin.

In return for profitable work, won on the basis of better value, the supplier may seek to drive a relationship with the client which does not rely on tendering every project. The costs saved in negotiating rather than tendering can be put to better use improving the current product, developing innovative processes and products or simply developing a greater understanding of the customer's business. Through better understanding of the customer's business comes a further opportunity to develop the relationship and win more work.

Before deciding which clients to pursue on a value basis, the supplier needs to consider their own value criteria. What are the key business drivers for suppliers' organisations? Over the years the key constructor, consultant and specialist objectives in initial partnering workshops have been identified as:

- profitability – sometimes defined as a reasonable margin or good cashflow
- do it once – right first time
- buildability – the opportunity to propose changes to the design and innovative ideas to add value not only for the client but also for the supplier
- certainty of work – in frameworks this would be a continuing programme
- support for other opportunities – particularly for single projects, the opportunity to use the project as publicity for winning future work
- local work (or projects in an appropriate location).

Each supplier should identify their own key value criteria, prioritising and weighting these in order to score, compare and prioritise potential clients. This process will help to ensure that the organisation does not chase unattractive or unwinnable contracts but targets a small number of key clients. In this chapter we will develop a client evaluation strategy using the six criteria above. We will also include one further criterion which we believe may be important to suppliers entering a partnering and integrated teamworking relationship – the willingness of the client to share risks, costs and rewards.

Before receiving or considering any bid documents for partnering and integrated teamworking projects, the supplier should set up a process to evaluate each current and potential future client against the supplier's value criteria. We suggest that one of the directors should take ownership of the process and pull together an internal team to carry out this evaluation. This supplier's client evaluation team should include representatives from all departments of the supplier's organisation. For example, a constructor may want to include estimating, design, surveying, finance, after sales, administration and marketing/client relations. We have included administration because we have found that the client's first impression of a supplier's organisation is probably through administration – at reception or on the telephone. Members of the administration department may also know the clients as thoroughly as the marketing department. In addition, it is important to have one's own administration team supportive of the evaluation process because the alignment of all departments' objectives is a reflection of good internal partnering.

The first step for the supplier's client evaluation team will be to set out the criteria against which each client will be evaluated. In a short session it should be possible for the team to arrive at a shortlist of around five to ten key evaluation criteria based on the supplier's previously identified value criteria. The evaluation criteria should then be weighted and our suggestion is that each member of the team allocates a percentage against each criterion. We will use the six key criteria that we identified above plus risk and reward. For seven criteria, the average weighting will be 14%. Any criterion that scores less than 5% may have minimal impact on the score and could either be discounted or rolled up into another category.

The supplier's client evaluation team should separately assess the weightings for each criterion. Once all have come to their own conclusions, the evaluation team should share their weightings and reach consensus (see Fig. 8.1).

Having determined the weightings, the supplier's client evaluation team should assess how they will rate each client against each criterion. As with the client's supplier selection process, we use a four-step evaluation of whether or not the team consider that the client has the appropriate culture for the supplier to enter into a partnering or integrated teamworking relationship. In this scale, an unacceptable performance scores zero, acceptable scores two and exceeding expectations scores three points. One point may be used for clients whose performance is currently unknown or who may be in the process of moving from price-based selection to value-based.

The supplier's client evaluation team now has:

- a set of their value criteria
- a schedule of evaluation criteria based on the supplier's value criteria
- weightings against each of the evaluation criteria
- a mechanism to compare and score clients.

During the next stage of the process, the supplier's client evaluation team will develop guidelines enabling them to make an objective assessment on how well specific clients meet the supplier's key value criteria. For example, within the risk and reward criterion, zero may be awarded for clients who are not prepared to share any benefits accruing from joint value management exercises, two points may be awarded to clients who have a joint risk management process in place and three points for clients who proactively promote and

Profitability	30, 40, 45, 30, 35	=	**36%**
Right first time	10, 15, 25, 10, 15	=	**15%**
Input to buildability	15, 0, 10, 25, 5	=	**11%**
Certainty of work programme	15, 10, 10, 5, 10	=	**10%**
Support for other opportunities	5, 5, 0, 10, 5	=	**5%**
Location of projects	10, 10, 0, 15, 10	=	**9%**
Risk and reward strategy	15, 20,10, 5, 20	=	**14%**

Figure 8.1 Client evaluation criteria weightings.

share the costs of partnering workshops and have a structured joint risk and reward share process in place.

After agreeing the weightings, the scoring mechanism and the guidelines, it is up to the individual evaluation team members to score each client in each criterion. This will result in a single score for each client as shown in Fig. 8.2.

Once the supplier's client evaluation team has rated and ranked all its prospective clients, the team can move to identifying preferred clients – those clients who have achieved the highest scores in the matrix. From this stage, the supplier can prioritise and pay more attention to the bid documentation for preferred clients.

A structured approach by suppliers to evaluating clients on the basis of better value to the supplier will initially assist the supplier's bid team in deciding whether or not to bid. If they decide to bid, the bid team may assess the depth and breadth of their responses to each individual question in the document rather than turning out the same CVs, case studies and radar charts that have been used on a series of previous projects. This attention to detail will help the bid team to produce a greater proportion of winning bids because they respond to the specific needs of the clients.

	Score		Weighted score
Profitability	**36%** x 2	=	72
Right first time	**15%** x 1	=	15
Input to buildability	**11%** x 1	=	11
Certainty of work programme	**10%** x 3	=	30
Support for other opportunities	**5%** x 3	=	15
Location of projects	**9%** x 3	=	27
Risk and reward strategy	**14%** x 1	=	14
	Total score	**=**	**184**
	out of a possible		**300**

Figure 8.2 Client evaluation criteria scores.

9 Submitting Successful Partnering Bids

Once the supplier has identified and prioritised their client list, it only remains for the supplier to convince the client and their advisers that their bid delivers better value than other bids. In a best value supplier selection process, that may not mean being the lowest price tender.

We have said that it is important for the client to set out their value selection criteria and weightings clearly in order to receive best value bids. It is just as important for suppliers to use this information to concentrate their efforts in the right areas to win the bid. For example, if sustainability is not one of the client's value criteria and the topic is not mentioned anywhere in their documentation, then there may be little value in the supplier's bid team expending resource on the topic. The bid team should concentrate their efforts on the client's value criteria that are highly weighted.

From experience as assessors, we find that it is the supplier's attention to presentation of a bid that first catches the eye. There is no doubt that a well-presented document will trigger a positive attitude in assessors.

We find it surprising how few organisations appear to do more than pull together standard corporate responses to specific client requests or questions. It seems that some bidders don't pay much attention to the actual question. They simply submit standard sheets with no reference to the specific client, project or bid. We have had examples where we have stated, 'Responses to any one question should be no longer than one A4 page in 10 point Arial type – any response longer than this will score zero points for that question.' Yet we have received documentation of two or more pages against a question, on one occasion with the comment, 'We find it difficult to fit our two page CVs with photograph onto one page and we trust you will find this in order.' The assessor's response to that should be to award zero points in accordance with the pre-circulated rules for suppliers and assessors. The supplier had read the instructions and decided not to adhere to them. Whilst the assessor's response with a zero score may seem a harsh decision, it is in line with all the rules set down and communicated to all bidders. For the assessor not to adhere to the rules would be unfair to other suppliers. In order to obtain maximum scores the supplier has to put in effort to meet the needs of the client.

We will not dwell on the price aspect of bids in this book except to say that each client and consultant may require their pricing to be calculated and presented in a specific way and the bid team must take account of this. Attention to detail is key. For example, some clients ask for overheads and profit submissions but do not state what should be in the overheads and profit. In this case, the supplier should ask for further detail of what is required and how the client will make comparison between bids that have different content. The supplier should continue to work at identifying and satisfying the client's prioritised value criteria. Suppliers cannot hope to meet the needs of clients if they do not clearly understand them.

It appears that many suppliers, at the bid stage, separate the price and quality documentation and send the separate parts to be worked up by different departments (for example, estimating and marketing). These departments may only come together in the adjudication meeting two or three days before submission. In our view, this is a missed opportunity for the supplier to build and demonstrate the value of internal partnering within their organisation. Regular and formal contact between the separate departments responsible for preparing the bid will enable the supplier team jointly to assess the risks and potential rewards and to identify innovative proposals that may gain additional points. To move from 'acceptable' to 'exceeding expectations' in any of the client's selection criteria may require as little as a single sentence in a proposal. For example, in response to a question on the supplier's record on training, a supplier may make a proactive commitment to sharing the costs of joint training. This solution may not have occurred to the client and could trigger the 'exceeding expectations' score. If that criterion is weighted at 20% of the quality scores, then this move from 'acceptable' to 'exceeding expectations' will improve the supplier's bid from 2 points to 3 multiplied by the weighting of 20. This is an extra 20 points out of a possible 300. A 7% uplift (20 points out of 300) in the quality score could make a significant difference in the final rankings.

Clients who are seeking to place contracts on best value will have spent time preparing their value selection criteria. These criteria and weightings should be set out in the documentation but, if they are not, the supplier should request this information in detail. If the client or their advisers are not prepared to open their books in this way then there may be a doubt about openness.

Suppliers should look for mention of an exit strategy in the client's selection documentation. Break points may be clearly set out (these may be based only on poor supplier performance) but the actual processes of determination may be less clear. An exit strategy should be agreed between suppliers and their prospective client partners before formalising the partnering arrangement and, if this is not clear in the selection documents, it should be proposed to clients by suppliers in partnering bids. Clarity and agreement on an exit strategy ensures that all partners are aware of the specific event(s) that will lead to a determination of the arrangement and the processes that will be followed by all partners in ensuring a determination without rancour. All partners should ensure that the exit strategy preserves the spirit of trust that marked the start of the relationship and ensures the ongoing maintenance of commercial confidentiality and intellectual property rights. Suppliers should be aware of (or should propose to the client) the exit strategy for each of the following breaks before the arrangement is set up:

1. expiry of the statutory period
2. consistent performance below agreed and clearly specified levels by one or other of the partners
3. an agreement by both partners that the arrangement must be determined.

Having identified the client's value criteria and weightings, the supplier bid team should prioritise their efforts on the higher weighted criteria, questions and sections of the documents. Addressing the criteria that are important to the client makes good business sense. It shows the client that the supplier understands and is prepared to address the client's needs. Clients and their advisers will be pleased to identify that suppliers have targeted their efforts on the client's key value criteria.

The supplier's bid team should ascertain whether the client is looking for innovative proposals for the project and the added value that will be delivered by integrated teamworking. The supplier may lose scoring opportunities if the bid contains, for example, an excess of case studies of previous experience (especially if these have not been requested) and insufficient proposals on the project. There may be potential, within the bid team, for one member to play the role of the client and check whether specific questions are being answered with the client's value criteria in mind.

The supplier should consider carefully how to respond to a question such as, 'How will you ensure the project does not go over budget?' In our opinion, this question has been phrased poorly in the context of an integrated team. It is very tempting to respond, 'By rigorous cost control of all specialists.' This may not be what the client is looking for and, in any case, flies in the face of the ethos of joint responsibility within an integrated team. Clients and suppliers should always, in a partnering and integrated teamworking context, be aware of the interdependency of all partners. Thus a more considered response to the question posed above might be, 'The integrated team of client, consultants, constructor and specialists will work together in an agreed programme of structured partnering, value and risk management workshops to identify ways in which the team can ensure budget certainty.' Note that this statement says how the value will be delivered (the partnering workshop programme) and identifies what value will be added by restating the client's value criterion (budget certainty).

Having set out their response to the client's value criteria and identified how they will deliver the added value, it may now be appropriate for the supplier to share their previous experience, either with the same client or with others as appropriate. However, this should be done bearing in mind the client's stated value selection criteria. It may be beneficial to consider rewording case studies to highlight the key points relevant to the client's values. This will increase the probability of a higher score in a specific selection criterion. If the case study does not look likely to add to the score, then the supplier should consider omitting the case study, as the additional irrelevant information will simply camouflage the more relevant points that are being made in the bid.

The full bid team should come together before the bid is finalised. This session is not only to check the pricing but to carry out a full review of the submission, quality and price. It may be helpful if one member of the supplier's bid team reads the client's questions, requests and instructions aloud to the bid team. In response, the full team should assess whether each and every section of the bid document accurately and concisely addresses the client's stated value criteria in the required format. The supplier's bid team should also ensure that the bid document clearly demonstrates the supplier's contribution to the project and identifies the benefits that will accrue.

Finally, the supplier should ensure that the presentation of the document is excellent. The team should consider (or find out) whether the client's assessors will have to unbind the document before copying or sharing around the team. For example, the sections on finance may be given to one assessor and the section on technical issues to another. This may influence how the document is assembled. Time spent on presentation will be time well spent as a well presented document will trigger a positive attitude in assessors. Once this positive attitude is established, the supplier's concentration on responding to the client's specific needs (answering the specific question) will add to the probability of a higher score.

10 Assessment, Evaluation and Award

In the final stage of the selection process, the client will appoint a team of assessors to evaluate the shortlisted suppliers' bids. The approach set out below for the final evaluation of the shortlist may apply equally to the evaluation of longlist responses, although less resources and time may be required due to the lower level of detail submitted.

On the day of the assessment and evaluation of the supplier bids, the assessors should meet early to review the scoring and assessment process, ironing out any potential misunderstandings and allaying any concerns regarding the processes and the timetable for the day(s). The facilitator should set up a spreadsheet scoring matrix before the meeting.

It is good practice for each assessor to be provided with a copy of each of the returned bids. This avoids the waste of assessors waiting for scripts to read and means that notes made on scripts by one assessor do not influence others. Also, if each assessor has their own copy of each bid, they are more easily able to refer back and make comparisons. In our experience, this outweighs the disadvantage of the additional copying and printing incurred.

When the bids are opened, each assessor should review each response for every supplier using the criteria weightings and the scoring matrix (for the purposes of this chapter we will assume the 0–3 process) to arrive at an auditable, objective, weighted score for each potential partner.

The facilitator should provide each assessor with a standard checklist of the criteria that has been previously agreed in order to assist assessors in being consistent with their scoring. For example, in

Chapter 7 we identified such criteria for off site manufacture (OSM) as follows:

0 No awareness of OSM or misunderstanding of manufacturing pressures
1 Awareness of the difficulties – 'we'll cross this bridge when we come to it'
2 Clear proposals for integrating prefabrication into the design and construction process
3 Already prefabricating successfully with quantified proof of benefit.

Copies of notes from visits to head offices and sites may be made available to the assessors if these are appropriate to one of the selection criteria. We have, for example, worked with clients who have made 'Can we work with them?' a selection criterion because they believed that this was essential to the integrated teamworking culture. In the event that the client's supplier selection team have not set up such a selection criterion then the submitted bids alone should be taken into account.

We find that a large boardroom table is suitable for the majority of assessment and evaluation sessions. These tables are generally large so each assessor can set out a reasonable quantity of documents to facilitate comparison of response A with response B, etc. Circulation space should also be provided so that assessors can move freely around and exchange documents as appropriate.

Each assessor scores each criterion for each supplier on the 0–3 scale previously set out by the client's supplier selection team. Assessors should not be permitted to give a higher score just because the assessor knows that the supplier has failed to answer the question to their best advantage. For example, we have heard some assessors say that a supplier is very good at one aspect of partnering but they haven't demonstrated it in the response. To maintain probity, assessor scoring should be solely on the basis of documented evidence, as it would be with a priced bill of quantities under a price-only competition.

Assessors should only score responses in whole numbers – the supplier has either achieved the rating or not. The ratings should have been set to enable the consistent interpretation of scores. If the assessor is in doubt, then it is probable that the supplier has failed to demonstrate achievement of the upper level and the assessor may credit the supplier with the lower score.

Some criteria may seek a number of responses. The assessors will evaluate each response on its merits and score it accordingly. At the end of that criterion, the assessor will review the scores given to each response and score the selection criterion overall in accordance with the 0–3 scale. Note that the assessor should only enter 0, 1, 2 or 3 and not an average of the responses as averages can hide serious failings in a supplier's responses. For example, Supplier A who scores 0, 3, 3 and 3 in a four-question criterion will score the same average (2.25) as Supplier B who scores 3, 2, 2 and 2. However, Supplier A has one unacceptable response whilst all the responses from Supplier B are acceptable or better. There are alternatives to dealing with multiple response scoring and the client's supplier selection team should select which alternative scoring method will be used before the evaluation and assessment session takes place. One option is for the assessor to use the lowest score against any response in scoring a criterion (thus Supplier A scores 0 and B scores 2) but, whichever method is used, the result should be to reward suppliers who do not have unacceptable responses.

We have not previously dwelt heavily on consideration of price in the selection process, not because we think it unimportant but because there are many ways of assessing the price content of returned bids. Apart from the consideration of the overall percentage that should be applied to price (for example, 60% quality, 40% price), the client's supplier selection team should have considered how they will score (on the 0–3 scale) the suppliers' submitted prices. There are many ways of doing this and we offer one option for selection teams to adopt or adapt as appropriate. We will assume that price has a weighting of 40% compared with the qualitative responses totalling 60%. The qualitative responses have a maximum of three points in each criterion so the maximum overall score for qualitative responses is 60 multiplied by 3 = 180 points. The price response also has a maximum of three points so the maximum overall score for the price response is 40 multiplied by 3 = 120 points.

For a single contract where the budget is known, the assessors could award points on a sliding scale. On the assumption that the budget is approximately correct, it is possible (when industry margins are in single figure percentages) that any price more than 10% below budget has an error or is an attempt to buy the contract, although the assessors will need to make their own judgement on this.

- ❑ Tenders exactly on budget could receive 80 points (being the equivalent of 'acceptable' and scoring two points multiplied by 40)
- ❑ Tenders 5% under budget could receive the maximum 120 points (being the equivalent of 'exceeding expectations' and scoring three points multiplied by 40)
- ❑ Tenders more than 10% over budget or more than 10% under budget could receive zero points.

Plotting a curve on the basis of those figures produces the graph shown in Fig. 10.1 and scores can now be read off for every price bid.

Alternatively, the assessors could split the bids into pre-agreed bands identifying which are acceptable, less than acceptable, unacceptable and exceeding expectations.

If, during the evaluation and assessment session, any of the assessors has a query or a comment about a specific bid they should refer it to the facilitator who should seek consensus from all assessors and record the decision to ensure fair dealing and transparency.

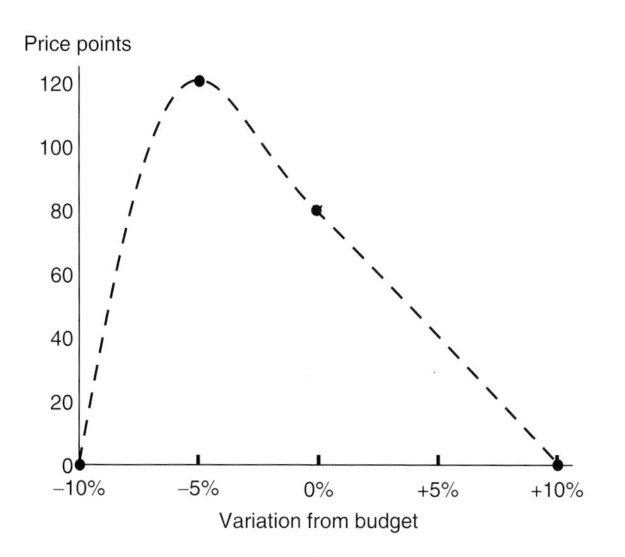

Figure 10.1 Sliding scale for price scoring.

Each assessor evaluates and scores each criterion for each supplier on the 0–3 scale and enters their scores on their personal checklist, provided by the facilitator (see Fig. 10.2). This maintains the probity of the process. Any scores that are subsequently changed, following group discussion and clarification, should be initialled by the assessor with reasons for the change.

On completion of all the scoring (on all criteria for all suppliers by all assessors) the facilitator should, regardless of the time of day, call a break. The assessors will have been concentrating on the responses for a considerable time and will benefit from 15 or 20 minutes break to clear their heads. In addition, there may be a need to refresh and tidy the room and set up the projection equipment for displaying the final selection matrix and scores.

Resuming the session, the assessors will share their scores for all suppliers for the first criterion. These should be displayed, either on flipchart or on projected computer display, in full view of all assessors and time should be allowed for them to discuss and consider the scores in this criterion. In the event that all scores for a supplier are the same, this score will be entered onto the final matrix as shown in Fig. 10.3. In the event that assessors' scores for a supplier are different, the assessors should discuss in detail why perceptions differ and attempt to come to consensus. Taking the average score across all assessors is one way of scoring responses but, as we have seen before, this takes no account of differences of opinion (e.g. a 0

Assessor___Name_________Supplier_________

Knowledge of sector	9%	3
Partnering credentials	16%	2
Safety and KPIs	11%	2
Environment and sustainability	7%	2
Value management	8%	1
Commitment to budgets	14%	2
Quality	12%	2
Supply chain management	6%	1
Off site manufacture	7%	1
Company profile	10%	2

Figure 10.2 Assessor's score sheet.

	Supplier			
	A	B	C	D
Knowledge of sector	3	2	2	2
Partnering credentials	2	2	2	2
Safety and KPIs	2	2	2	2
Environment and sustainability	2	1	3	1
Value management				
Commitment to budgets				
Quality				
Supply chain management				
Off site manufacture				
Company profile				

Figure 10.3 Assessors' consensus scores for all suppliers.

and a 3 scores the same average as a 1 and a 2 although one response identified the supplier as unacceptable). Where possible, the consensus view of the assessors should be entered on the final matrix (see Fig. 10.3).

In the final stage of the assessment and evaluation meeting, the facilitator will enter all consensus scores for each supplier in each criterion, multiply each supplier's score by the weighting for that criterion and total the extended scores for each supplier (see Fig. 10.4). This will result in an objective, auditable score for each supplier

		A	B	C	D
Knowledge of sector	9%	3=27	2=18	2=18	2=18
Partnering credentials	16%	2=32	2=32	2=32	2=32
Safety and KPIs	11%	2=22	2=22	2=22	2=22
Environment and sustainability	7%	2=14	1=7	3=21	1=7
Value management	8%	1=8	0=0	2=16	2=16
Commitment to budgets	14%	2=28	2=28	3=42	1=14
Quality	12%	2=24	2=24	2=24	2=24
Supply chain management	6%	1=6	1=6	1=6	3=18
Off site manufacture	7%	1=7	2=14	2=14	0=0
Company profile	10%	2=20	1=10	2=20	3=30
	Totals	188	161	215	181

Figure 10.4 Final extended totals for all suppliers.

and, on the basis that the criteria and weightings were based on the client's stated value criteria, will provide a best value ranking for each supplier.

After checking the calculations, the assessors can make their recommendations to the decision makers within the client organisation. We suggest that the decision makers attend the final session in which the scores are being finalised in order that they can see for themselves the transparent and auditable process that has been followed.

The decision makers should then inform the selected supplier (or suppliers in the case of frameworks and multiple appointments). They should offer the selected supplier the opportunity to join the client in forming an integrated team as full partners and to take part in the initial partnering workshop to determine mutual objectives, issue resolution processes and targets for continuous improvement to add value for all partners. After this offer has been accepted, a representative of the assessors or the client's supplier selection team should write to the unselected suppliers, promptly informing them of the situation and offering to share with them the results of the total selection process and an explanation of the reasons why they did not succeed.

11 Developing the Integrated Team

In the previous chapters we have shown how the culture change in the UK construction industry is driving a supplier selection process based on clients' perceptions of overall value to their business/organisation rather than a selection process based solely on lowest price. We have also shown how suppliers are making value judgements, evaluating which clients they prefer to work for, rather than simply cutting prices to attempt to win every job that is put their way.

In the earlier model of the industry, prior to the introduction of integrated teamworking, each organisation would, after contracting to carry out the project, withdraw to its technical and cultural silo to safeguard its own interests. Each organisation would be secure in the knowledge that all work had been specified, all costs agreed and the programme finalised. Yet too many clients, according to the executive summary of *Rethinking construction* (Egan, 1998), are dissatisfied with the overall performance of this earlier model of the industry. We suggest that this dissatisfaction is mainly the result of the retreat by all organisations into their technical and cultural silos. The relationship in which both sides (the client side and the supplier side) only look after their own interests, ensures that objectives are not aligned. In this disintegrated team model, the client looks to control its expenditure whilst maintaining the quality specified and the suppliers look to optimise their margins within contracted prices by reducing their costs in whichever way they find possible. This approach leaves little incentive for the client side to input further as they insist that the specified quality will be delivered at the contracted price. It leaves little motivation for the supply side to fund innovation or training as their margins are low.

We appreciate that not all previous contracts or relationships suffered from the silo mentality. Many teams tell us that they have worked cooperatively with their clients, consultants or constructors for many years. The wider industry should learn from the positive experiences of these organisations and also from those who have more recently successfully adopted the partnering or integrated teamworking approach.

The value based selection and evaluation processes that we have set out in the earlier chapters have brought together a full complement (not yet a team) of organisations prepared to work together for the good of each other and the project. Integrated teamworking does not allow individual organisations or their members to withdraw into their technical and cultural silos. There is no hiding place because there is a need to develop the combined intellect, skills and synergy of the team to deliver added value for the benefit of the project and all partners.

Accelerating change (Egan, 2002) set a target for 50% of construction projects (by value) to be undertaken by integrated teams by the end of 2007. It committed the Strategic Forum for Construction to produce an integration toolkit to help the industry to achieve this target and this toolkit is now available to all at www.strategicforum.org.uk. This website defines an integrated project team as:

- a single team focused on a common set of goals and objectives delivering benefit for all concerned
- a team so seamless, that it appears to operate as if it were a company in its own right
- a team, with no apparent boundaries, in which all the members have the same opportunity to contribute and all the skills and capabilities on offer can be utilised to maximum effect.

The website www.integrationsupportnetwork.org.uk lists appropriate practitioners who will assist organisations in their search for excellence in integrated construction teamworking.

The benefits of an integrated team, compared with a traditional supply chain, include:

- Teams work together but a chain is only as strong as its weakest link – integration of organisations leads to fewer links in the chain and consequently less weak points.

- Direct communication between team members reduces the possibility of distortion of the message through multiple handling along the chain.
- A team will work to support and develop a weaker member – this is more cost effective than re-tendering when one team member is struggling.
- As the team shares its knowledge, it will develop a bond of trust which will lead to more learning.
- Team learning will help individual team members in their personal development.
- Shared learning, shared knowledge and shared understanding encourages, enables and supports better communication.

However, development of this integrated team will require full commitment from all team members (organisations and individuals) and a major cultural shift for some. The two approaches are illustrated in Fig. 11.1. It will be necessary to integrate the team not only in

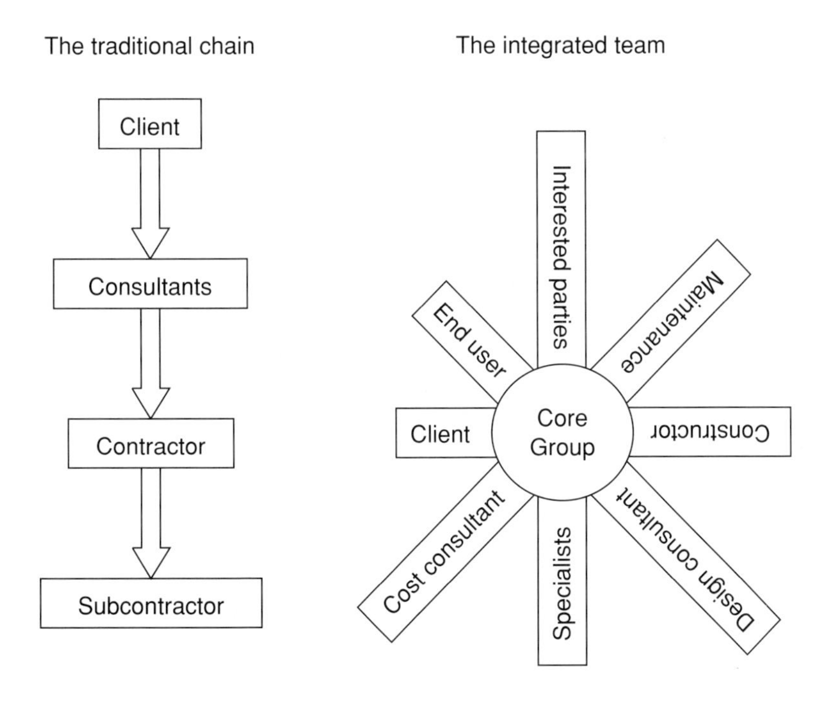

Figure 11.1 Supply chain or integrated team?

the regular technical design and construction meetings but also in training and team building workshops. This step will incur costs, both in terms of cash outlay and resource commitment, yet the added value from the integrated team will not be realised without this additional commitment.

Training in new technical or process skills will almost certainly have to be supplemented by training in teamworking skills. There is a need to bridge the cultural silos that may have provided our comfort zones for many years and for the whole team to accept that a project cannot be delivered without trust and cooperation from all involved. For example, building a successful sports team is not achieved simply by buying a squad of the best players in the world. This squad requires quality coaching and training to develop a culture of working together in harmony for the benefit of the team. Commitment to training and development away from the competitive arena is the way in which successful sports teams develop. Construction teams who wish to emulate the success of top sports teams must also understand the need for off-the-job training in order to identify and communicate common goals and develop appropriate strategies to reach them. Team bosses (senior management from all organisations) must also motivate their teams and instil in them a desire to succeed through encouragement, incentivisation, empowerment, praise and reward for good performance.

In early 2004 we facilitated a value management and team building day for the integrated client, design and construction team responsible for the Administration and Student Services Building for the University of Southampton.

Prior to the workshop, the University of Southampton's project manager, John Brightwell, emailed all team members as follows:

'On Tuesday 6 April you are invited to attend this event, which is intended to inject enthusiasm and a greater depth and breadth of understanding of the project in all who attend. To ensure the team is given every chance to gel, the day winds down with an evening meal, which is optional, but should be seen as a good way to cement relationships and encourage networking and the partnering ethos. The Treetops Restaurant has a good reputation and an attractive menu, and the grounds offer a chance of fresh air and a leg-stretch after the formal working sessions are over. Dinner starts at 7.00pm, and I need to know, preferably by return email, and in any case as soon as possible, how many are staying for dinner. I hope you will all be able to respond quickly, as I need to book table(s) for dinner, and hope that you will all go the extra mile, to make this a landmark event in launching the project.'

The informal evening dinner was attended by the whole team representing the client (University of Southampton), constructors Bluestone, consultants Nicholas Hare Architects, Northcroft, Hoare Lea and Anthony Ward Partnership and specialists Resource Environmental Services and VHB. All contributed, not only to the funding but also to the camaraderie of the event.

The importance of social interaction within the team should not be underestimated. The human interaction generated by social events enables the team to bond. Whilst the value of social events may not be fully appreciated by some in the team, these off-the-job sessions will focus on breaking down barriers, increasing mutual respect and creating a team that is pulling in one direction. An effective team that has a common focus will deliver much higher value than team members could deliver individually.

12 Trust

Trust is elusive, fragile and generally difficult to quantify but it is critical to the success of a partnering or integrated teamworking relationship. Teams know when they have developed trust and they know when they've lost it. But how is trust measured? How do teams build it up and maintain the high levels required to sustain a strong integrated team?

We've seen trust defined in an arithmetical formula by Dr Tom Sant (www.santcorp.com) and this has helped us to determine strategies for building up trust in integrated teams. Dr Sant's formula is:

'Trust = Positive Experiences divided by Risk'.

Positive experiences may include the demonstration of mutual respect, fewer defects, more projects completing on time and to budget, increased resident satisfaction, payment on time and sharing of rewards. Using the formula above, trust will increase for the same level of risk when the team identifies and communicates such positive experiences. As trust increases within the team, the members will be more open and honest with each other and this openness will enable them to jointly identify, assess, plan and manage risks more effectively. The consequent reduction in overall risk will also, using the formula above, result in a higher level of trust. This is illustrated in Fig. 12.1.

We identified above that trust is fragile. A trusting relationship that has taken years to build can be shattered by a single thoughtless or deliberate act or by one member of the team not doing what they

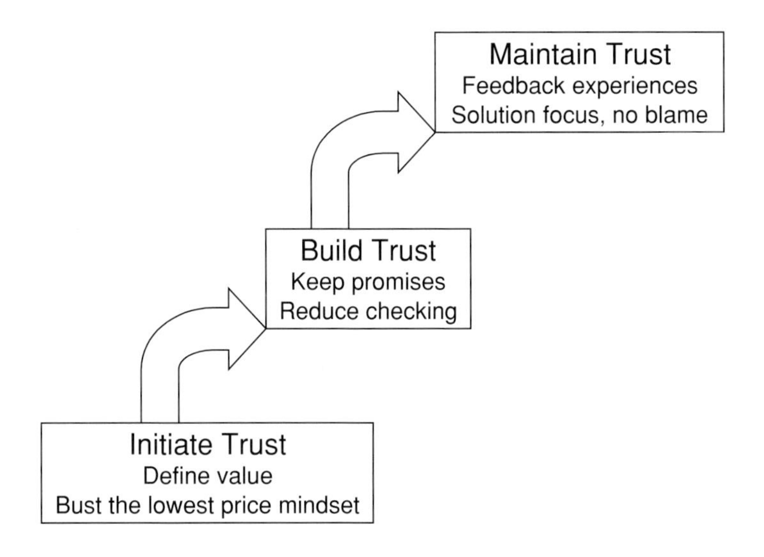

Figure 12.1 Trust.

said they would do. This single, negative experience may return the team to a position of trust substantially below their starting point and it may require a much greater effort on the part of the team members to rebuild trust, even to its initial level. Therefore, it is important that team members, core groups and senior management of the partnering organisations do all in their power to remove from the team any element – process or person – that is likely to undermine trust.

The Construction Industry Institute (http://construction-institute.org) highlights the importance of trust in team building as follows:

> 'One of the basic elements of a cooperative relationship is mutual trust. Trust is a salient factor in determining the effectiveness of many relationships. It facilitates interpersonal acceptance and openness of expression. A trusting relationship between the parties is based on a mutual understanding of each other's capabilities and limitations. It is also based on the personal and corporate integrity of both parties. Trust is a basic ingredient in effective team-building, timely decision-making, and in building long-term relationships.' (Construction Industry Institute, undated)

INITIATING TRUST

High levels of trust may not exist at the outset of a partnering or integrated team relationship. Consequently, if two or more organisations are to make the leap of faith from lowest price contracting to the delivery of best value through integrated teamworking, they should jointly demonstrate their commitment to working together in such a way that trust within the team can be built – for example, by joint training, ensuring that a common definition of value is communicated across the team, working towards open-book accounting or by co-locating team members so that cooperative relationships can develop more easily through face-to-face communication.

Before any organisation enters into a partnering or integrated teamworking relationship, they should identify whether they trust their potential partners. If an organisation were to contract with another organisation that they did not trust (however attractive their price point or programme of work) one should question whether they were doing the best for their shareholders, board members or elected members. Could it even be asked whether they were acting improperly or negligently in appointing a supplier that they didn't trust?

Trust depends very much on the interaction of individuals and on interpersonal relationships, yet it tends to be organisations, not individuals, that are appointed to teams. Some people in the industry cannot, or do not want to trust their team colleagues. They may have long memories of confrontational and adversarial relationships or they may consider that an adversarial attitude serves them or their organisation better. Despite this unpromising start, trust can be built as teams work together. Senior management, represented by the core group, should discourage the perpetuation of old adversarial and lowest price attitudes and encourage the early quantification and publication of positive experiences, especially where these arise as a result of team members working together across organisational boundaries.

For a team to be fully integrated, all parties need to trust and understand each other. In recognition and reinforcement of this approach, clause 1 of PPC2000 (The Association of Consultant Architects Ltd & Trowers & Hamlins, 2000) specifically requires its signatories, and by inference the employees of those signatories, to work in a 'spirit of trust, fairness and mutual cooperation for the benefit of the project'. Our experience is that a spirit of trust reduces waste, including the waste associated with low morale, defensive behaviour, checking and double-checking, writing contractual letters, preparing claims and fighting unwinnable battles. A cooperative approach from one partnering team member tends to breed a co-operative approach in another in return – adding to the positive experiences of the whole team and thus building trust.

BUILDING TRUST

A relationship in which trust is absent requires control. Control requires the application of considerable resource and this is a costly option. If, from a client's perspective, it can be proved that the supplier has delivered to programme and required specification, then the extent of the client's checking operation can be reduced. If, from a supplier's perspective, the consultant has delivered the design information in the appropriate form and the client makes interim payments on time, then the extent of the supplier's need to chase and check can be reduced.

A reduction in the need to chase or check immediately saves resource for the partners and builds trust within the workforce. Clients who do not trust their suppliers might think that a less rigorous checking regime could lead to a supplier taking shortcuts or installing sub-specification products. If this were the case, the client could revert to a full checking regime, fully aware of the attendant resource implications of this step. But would a supplier deliberately install sub-specification products and risk losing trust when there is so much more to be gained by installing to specification, building trust and reducing the costs of confrontation and checking?

If the client feels that they can reduce their checking regime on a supplier, this should be communicated to the full integrated team as a positive experience. This additional positive experience further increases the level of trust in the integrated team. Increased trust is likely to increase the motivation of the supplier's workforce and they are more likely to continue to deliver to programme and to required specification. This further positive experience should lead to further increased trust and an even further reduction in checking, driving yet further resource reduction on the part of the client and of the supplier. Thus it can be seen that trust, once built, can be self-maintaining.

The resources saved as a result of increased trust in the team may be reinvested in, for example, joint training and integrated team workshops to develop more innovative and value-adding processes, leading to the elusive win–win solution.

MAINTAINING TRUST

We believe that many individuals implicitly trust those with whom they contract until they are let down. In fact, many client and supplier organisations tell us that they have worked collaboratively for years without calling their relationships partnering, alliances or integrated teams. Trust has clearly been initiated, built and maintained in such relationships.

In order to initiate, build and maintain trust in the integrated team, the core group should monitor behaviours and individual partnering team members should flag up and address any issue that risks breaking the trust.

Trust is the basis for honest feedback and open exchange of information. If there is a lack of trust, one partner is unlikely to share with another the fact that the project has an issue. They will be afraid of the inherent blame culture. The partner's lack of willingness to share such information will prevent a team from addressing the problem and jointly seeking alternative solutions. This has the potential to exacerbate the situation, raising negative experiences and further reducing trust. If, however, the partners trust each other and flag up the issue, in the certain knowledge that blame will not be attributed but a solution sought, then the creative effort of the team can be put to seeking alternative options and trust can be built and maintained.

In the Lukely Court project that we facilitated for the Isle of Wight Housing Association the window supplier, Coastline Windows, hit supply problems in the first phase. The team could have taken the view that Coastline were in delay and changed supplier. Coastline, however, were solution focused and trusted the team to support them. They attended the continuous improvement workshop with proposals for changing their procurement strategy, overcoming the problem and clawing back programme delay at no additional cost to the client or constructor.

Teams and team members who trust each other add value by keeping their promises (reducing the need for one-on-one marking and checking) and feeding back their experiences in a no-blame, solution-focused culture. This greater reliance on each other's honesty and openness creates an environment where breaking that trust is less likely. As Brian Fox, managing director of SDC Builders, said in *Building* magazine, 'I believe that if someone trusts me then that places a greater onus and responsibility on me than any contract ever written' (Building, 1994).

13 Respect for People

If an organisation demonstrates respect for its staff then quality staff will beat a path to its door. Organisations that demonstrate respect for their workforce and address their issues tend to have less problems both recruiting and retaining quality staff. This demonstration of respect is key to the implementation of successful partnering and integrated teamworking.

Successful partnering and integrated teamworking relies heavily on the ability of team members to support and encourage each other. This applies not only to relationships across organisations but also to relationships within individual organisations. For example, each member of the team should be prepared to give time to respect and listen to other team members.

> J. W. 'Bill' Marriott, son of the founder of the Marriott International hotel and hospitality corporation, recalls that, 'My father, J. Willard Marriott, kept his executive staff waiting on many occasions while he sat on a hotel lobby sofa counselling a housekeeper or cook about a family or work problem. Far from being a waste of time, he considered such chats an investment in his company's future. He knew that a troubled employee couldn't deliver top-notch customer service. Simply by taking time to listen, Dad found himself surrounded by employees willing to put 110% effort on the job. The pay-off was tremendous: happier employees, satisfied customers and a successful company' (Marriott, undated).

In order for an organisation to assess its performance in its respect for people, Constructing Excellence has developed and published a set of key performance indicators (Respect for People indicators at www.kpizone.com). Using these indicators will enable an organisation to measure its status and progress in up to ten key performance indicators:

1. employee satisfaction
2. staff turnover
3. sickness absence
4. safety
5. working hours
6. qualifications and skills
7. equality and diversity
8. training
9. pay
10. investors in people.

A standard questionnaire and supporting formulae have been developed to assess an organisation's performance and are provided in the handbook. Constructing Excellence also publishes annual data on the standards achieved across the construction industry so that individual organisations can track not only their own progress but also their progress against their peers. A further twenty-one secondary performance indicators (any of which may be used as alternatives to any of the ten KPIs) are shown in the Respect for People KPI handbook (Constructing Excellence, 2004), complete with formulae and annual performance measures. We do not recommend that an organisation or integrated team regularly measures more than about ten indicators as measuring too many indicators will consume considerable extra resource for little additional benefit.

Attention to the cultural aspects of an organisation, and of partnering or integrated teams, will add value in many ways. For example, greater employee satisfaction (also reflected in lower staff turnover and sickness absence) will increase the productivity of the team. Shorter working hours, combined with appropriate skills and safety training, should lead to a lower incidence of reportable accidents. These initiatives and the ability and willingness of the organisation to pay appropriate salaries and wages should lead to a more motivated, satisfied and loyal workforce and thus to lower staff turnover and increased continuity of personnel within the team.

Training is essential to development of the integrated team. The cost of joint training in technical, process and teamworking skills is likely to be less than the potential cost of low productivity, a poor health and safety record and replacement of unmotivated team members. The cost of recruiting and training a new member of the team has been measured at around 20% of the annual salary of the employee and this does not take into account the disruption to the team members of either the unmotivated leaver or the new employee.

Attractive and appropriate (note that we have not said *adequate*) site facilities will also attract quality staff. In our experience, there is a correlation between the quality of site facilities and the effectiveness of project teams. We visited one site recently where the organisation was proud of the fact that they had a ladies toilet. However, this was locked (so that the men didn't use or abuse the ladies' facilities) and if any female wanted to use the toilet they were asked to obtain the key from the site agent. We wonder what message this gives any visitor or prospective team member.

Applying equality and diversity in the workplace is about ensuring that all team members recognise the diverse nature of each individual's needs and their technical and non-technical skills. The team should make best use of these skills and develop them further for the joint benefit of the team and the individual, whatever their gender, ethnicity, disability, age, background, personality or work style. This will tend to reduce turnover as individuals feel valued for their contributions to the team. As a result of addressing equality and diversity issues, teams will find that their organisations are more attractive to work for. The positive image presented by the team will increase the potential to attract team members from a much greater proportion of the community.

A number of organisations exist to support constructors and clients in the private and public sectors to measure and develop their approach to building a trained, efficient and motivated workforce, leading to more productive teams delivering higher value projects. Two such industry-specific organisations are the Construction Clients Charter and the Considerate Constructors Scheme. These are supplemented by Investors in People, an organisation that works across all industries.

tion clear. Has it been sent to the appropriate people? Sending to too many, just in case, is as improper as sending to too few.

- There may be a lack of consideration of how the message will be received. Can the email be misinterpreted? Email does not convey intonation. This is why there is major potential for misunderstandings over email communication.
- Team members should be aware that, whilst the writer has a responsibility to deliver the message carefully to ensure comprehension, the receiver also has a responsibility to attempt to understand the intent in the message and not to take umbrage where no offence is meant.
- If the email is urgent, will a phone call be better?

Facial expression and body language are major factors in good open communication. It is important to ensure that facial expression matches the meaning of the spoken word and to be aware of the emotion that our facial expression is portraying. For example, is the facial expression one of interest or boredom, joy or sorrow, anger or calm, anxiety or relief? Facial expressions in the speaker may be mirrored by the person receiving the message. For example, it is likely that a lively, energetic proposal will receive a lively, energetic response. In some team meetings we have found that team members may fail to get their message across as they contribute with little or no voice projection and no energy, whereas those who use a 60db voice and a positive body posture communicate effectively.

Trust is reflected in unfolded arms, leaning forward with interest when another is speaking, encouraging nods, a smile when appropriate and good eye contact. Eye contact is critical, not only to getting the message across but also in ensuring that the message is being received by the other person. Conversely, breaking eye contact sends the message that the conversation is at an end.

Whilst body language is key to good communication, the ability to listen effectively is also a major element of communication. Yet, in our experience, it is a rare skill. We developed a listening exercise (set out in Chapter 38) following a comment at Unite's first annual partnering review in January 2004. David Livingstone, then Unite's Development Director, identified that:

> 'People here are experienced, tough, technically competent and confident but we may have pre-formed views. There is a danger that we think that

what we say is more important than listening to you. We can't listen when we talk. We need the self-awareness to know when to shut up and listen at a really deep level to what you are saying. All of us need to focus on listening so together we can find a better solution.'

A solution-focused team is one where there is no blame culture. The development of a no-blame culture depends on open and honest communication in which the parties accept their responsibilities and are not afraid to flag up issues for resolution at an early stage. The team members accept that they are all in it together and that any problem impacts on them all. The attitude should then be, 'Let's get on with finding the solution rather than escalating this into a dispute.' In this environment, team members should feel confident in the support of the team when they openly tell their colleagues they are in a difficult situation. The team member with a problem should know that this openness will not provoke the automatic response of, 'What have you done?' or, 'Why did you do that?' but a solution focused response such as 'How do we get out of this situation?'

Many difficult situations for partnering and integrated teams can be avoided by acting reasonably and without delay. Delay costs time and money and builds distrust so partnering-specific contracts put the responsibility on team members to act reasonably and without delay, reducing frustration for other team members and increasing the efficiency of the team. Such timeliness is made easier if the partnering team members trust each other and have been encouraged to talk openly and work cooperatively. They are then more likely to seek information by picking up the phone or emailing colleagues rather than writing contractual letters.

In order for team members to respond reasonably and without delay the person making the request should be specific about the timescale for the response and resolution. For example, 'I need a response, please, by 5pm on Friday 1 August and resolution of the issue by Friday 8 August'. In some cases it is not possible to give an immediate response to a question. So, the person to whom the query has been addressed should respond without delay with a reasonable comment such as 'Thanks for your query. I don't have an answer today but I should be able to give you a response within 48 hours.' This keeps the communication channel open and reduces frustration.

As trusting relationships develop within the team, communication will reach a higher level. This may not develop as far as extra sensory perception but it is likely that team members will be able to anticipate each other's needs and requests for information, particularly in term contracts and frameworks where there is an element of repeat processing.

Communication is a huge issue and is key to the efficiency and effectiveness of partnering and integrated teams. Whilst early and efficient dissemination of information such as roles and responsibilities, issue resolution processes and workshop programmes is a sound basis for communication around the team, a raised awareness of the significance of intonation, body language, effective listening and the potential for misunderstandings will enable team members to be positive, assertive and supportive in their communication. This will build trust and increase the efficiency of the team.

15 Non-technical Team Roles

It is no coincidence that Sir Michael Latham's 1994 report was titled *Constructing the team* (Latham, 1994) or that the subsequent best practice guide to construction partnering was called *Trusting the team* (Bennett & Jayes, 1995). A team that works well together produces far more than the sum of its individual parts but all organisations should be aware that this does not happen accidentally or as a direct consequence of working together for some time. From the first time that the core group or partnering champions meet they should be considering how to develop the team culture so that all team members support and develop each others' skills for the benefit of the project. It is essential that the core group keep team building in the forefront of their thoughts and plans as a poor team atmosphere will build mistrust and drive inefficiencies.

Integrated teams in construction tend to be made up of individuals who work for organisations which have been selected on the basis of the past record of the organisation, not necessarily the individual's skills or personality. Even in the rare case that specific individuals have been chosen to be part of the team, it is likely that they will have been chosen for their technical skills rather than their ability to perform within the integrated team environment. It is, therefore, the core group's responsibility to focus the team, developing individuals' strengths and addressing their weaknesses for the overall benefit of the team and the project. Nevertheless, despite the unifying role of the core group, it is still the individual team member's responsibility to adapt their own personal and professional attributes for the overall benefit of the team.

We have probably all seen geese flying in a V-formation but we may not have appreciated the rationale – the goose's business case – for this:

- As each goose flaps its wings, it creates an uplift for the birds that follow. By flying in a 'V' formation, the whole flock adds 71% greater flying range than if each bird flew alone. In the same way, team members whose objectives are aligned and who possess a strong sense of teamworking, have greater potential because they are supported by others.
- When a goose falls out of formation, it feels the drag and resistance of flying alone. It quickly moves back into formation to take advantage of the lifting power of the bird immediately in front of it. Thus, stepping out of line makes life harder for team members. Staying in formation with those headed where we want to go helps us and them but we have to be willing to accept their help and give our help to others.
- When the lead goose tires, it rotates back into the formation and another goose flies to the lead position. Team members are also interdependent on each other's skills and capabilities. It pays to delegate leadership where appropriate – for example, leading topic-specific task groups.
- The geese flying in formation honk to encourage those up front to keep up their speed. In teams where there is encouragement (whether from the leaders or from the supporting members), the effectiveness of the team is much greater.
- When a goose gets sick, wounded, or shot down, two geese drop out of formation and follow it down to help protect it. They stay with it until it dies or is able to fly again. Then, they launch out with another formation or catch up with the flock. Geese in formation, therefore, have an exit strategy which does not consist of dumping one partner when times get tough. There are longer term benefits in standing by each other in difficult times as well as good.

A focus on interpersonal skills and relationships is critical in building successful integrated teams. Whilst individuals may be exceptionally technically competent, there will be no benefit to the team if they bring an attitude to the project that runs counter to the team spirit that the core group is building. One bad apple will spoil the barrel. Accommodating those who do not want to fit in to a team (or who are unable to do so) is wasteful of resource and may run the risk of alienating those who are committed to partnering and integrated teamworking. In our experience, where the core group has dealt swiftly and positively to address interpersonal issues, the results from the team have improved substantially. We recommend that the core group identifies and deals with such cases, addressing any problems through frank one-on-one discussion, offering the individual the opportunity to change, supported by retraining where appropriate. In extreme cases, after discussion and retraining, the core group should consider removing from the team any individual with anti-team behaviour.

We have found that using the Belbin Team Role methodology with the full team of client, consultants, constructor, specialists and end-user, raises the team's awareness of each individual's non-technical strengths and weaknesses. Individuals' responses to simple standard questionnaires produce individual and team reports generated by the Interplace expert system. Sharing these reports in a team environment (for example at the initial partnering workshop) enables the team members to be aware of their colleagues' strengths and weaknesses. This understanding of non-technical skills and personality types helps create and build successful integrated teams.

Dr Meredith Belbin (Belbin, 1981) identified eight non-technical team roles and, subsequently, the additional role of specialist. All of these non-technical team roles are essential in developing a balanced team and no one role is more important than any other. It should be noted that, within a team, each member may fulfil more than one role and this is identified in the reports generated through the Belbin Interplace expert system. The strengths of each team role are as follows:

Plant – creative, imaginative, unorthodox; solves difficult problems

Resource investigator – extrovert, enthusiastic, communicative; explores opportunities; develops contacts

Coordinator – mature, confident, a good chairperson; clarifies goals, promotes decision making, delegates well

Shaper – challenging, dynamic, thrives on pressure; drive and courage to overcome obstacles

Monitor evaluator – sober, strategic and discerning; sees all options; judges accurately

Teamworker – co-operative, mild, perceptive and diplomatic; listens, builds, averts friction
Implementer – disciplined, reliable, conservative and efficient; turns ideas into practical actions
Completer – painstaking, conscientious, anxious; searches out errors and omissions; delivers on time
Specialist – single minded, self starting, dedicated; provides knowledge and skills in rare supply.

We recently conducted a Belbin Team Role analysis on a nine-member team. In addition to producing self-perception reports for each team member, the software generated a team report which identified that,

'This team contains several members with restless energy who like to get things moving. They may become frustrated in slow moving situations. Unless the pace quickens, interest is liable to flag. For this reason any meetings should be lively, the agenda should be kept short, and crisp, clear decisions should follow any debate. The team as a whole is more geared to seizing opportunities than to detailed planning. Particular individuals will need to be brought in at the right time if the team is to become fully effective.'

The report went on to identify who should be consulted for new lines of thought, exploitation of opportunities, coordination of group effort, increasing the pace, choosing between options, improving team atmosphere, turning decisions into workable procedures and completing plans without mishaps. This report helped the team leaders to recognise that there were gaps in its armoury and to consider whether to bring in new team members to bolster the planning phases of projects.

Project leaders and core groups who build their teams only concentrating on technical skills tend to perpetuate the silo culture. Those who also acknowledge and consider the importance of non-technical skills in building an integrated team, add value not only to the project but also to the personal development of each team member.

16 Mutual Objectives

Mutual objectives, the alignment of corporate and individual goals, is the first of the three essential features of successful partnering arrangements identified in *Trusting the team* (Bennett & Jayes, 1995). Without alignment of objectives, members of a team will be pulling in different directions and potentially cancelling out each others' efforts by solely concentrating on their own objectives. The purpose of developing a set of mutual objectives is to harness the power of the whole team and focus their efforts on pulling in the same direction. Mutual objectives of an integrated partnering team are set out in a partnering or team charter and we have set out our recommended process for developing such a charter later in this chapter.

We note that some partnering charters have been written by the partnering consultant, facilitator or adviser from lists of objectives submitted by team members prior to the initial partnering workshop. Whilst this may speed the day, we recommend that this practice is not adopted as there is a danger that this approach may lead to lists of organisations' selfish objectives which do not necessarily concur with, refer to or consider the objectives of the other organisations. These lists do not overcome the potential for the team pulling in different directions. In fact, they may reinforce it. Obtaining mutual understanding and consensus through discussion and full team input to the process of developing a partnering charter of mutual objectives is necessary in order to obtain buy-in from the team members for the duration of the partnering relationship. Moreover, there is likely to be a higher level of buy-in to a charter where the team has developed the wording than to a charter where the wording has been developed by a third party.

Defining and committing to the mutual objectives of the integrated team should be the first task for the integrated team within the initial partnering workshop, after introductions, objectives of the day and an icebreaker to relax the team. During this first stage of the workshop, the team will communicate and identify the value criteria and the objectives of all partners. Through an iterative process, the team will work these into a statement of mutual objectives in the form of a signed non-contractual partnering or team charter.

Senior members of the organisations need to attend the initial partnering workshop in order to present their organisations' objectives, demonstrate commitment, guide their team members and sign the charter. Other interested parties and members of the integrated team who should be encouraged to be present at the agreement of the partnering charter of mutual objectives include end-users (including residents and workforce), auditors, best value managers, finance managers, key specialist contractors, consultants and the decision makers from all participating organisations.

An optimum number of delegates for an initial partnering workshop is between 15 and 25. Larger teams may prove rather unwieldy for the group tasks and smaller teams may be unrepresentative of the full partnering team. Except in full team sessions, we have found that it is effective to keep group sizes to no more than seven in order to stimulate discussion and maximise individual participation. Note that we use the term *team* to denote the full complement of attendees and *group* to denote smaller subdivisions of the team.

Initially, the facilitator should divide the team into a number of organisation or role-specific groups (for example five separate groups for clients, consultants, constructors, specialists and interested parties) and provide each separate group with a flipchart pad and a different colour flipchart pen. There is no particular significance in the different colours. This simply helps the facilitator to identify the different sheets in the feedback session and when writing the report.

Each group should be tasked with identifying their five key selfish objectives and recording these on a flip chart. In our experience, 15 minutes should be sufficient for this but the facilitator should be flexible in timing to allow groups to fully understand the issues and clearly formulate their objectives. Depending on the venue, it may be beneficial for some groups to find a breakout area outside the main room to carry out this task. The group's objectives should be clearly set out on the flip chart and should be comprehensible without further explanation. A single word (e.g. 'cost') is rarely sufficient and the facilitator should circulate amongst the groups to prompt the members to be more specific in their objectives.

At the end of this group session, the facilitator will reconvene the team and ask each group in turn to display and feed back their objectives to the full workshop team. Five minutes should be allowed for each group. The facilitator should number the points consecutively in the feedback session (e.g. 1 to 5 for group 1, 6 to 10 for group 2, etc.). This assists the team and the facilitator in referencing individual objectives during the later sessions. Team questions, comment and discussion on the selfish objectives should be encouraged so that full understanding is reached. A good deal of commonality is usually identified at this point and it is important that the facilitator elicits this. We (and the teams) find it useful in this stage if the second facilitator enters the numbered objectives onto computer, printing multiple copies of a single sheet of the selfish objectives for groups for reference in the next session.

In the second group session, the facilitator will divide the team into a number of cross-organisational groups with a maximum of seven members in each (we will assume four groups of seven in this example). As far as possible, each group should contain representation of client, consultant, constructor, specialist, interested parties, etc. The facilitator will hand each group two or three copies of the single sheet of the selfish objectives for reference during this session. Each group should first consider and discuss all the selfish objectives. The groups should be given up to 20 minutes to compile 5 statements of joint objectives that take into account all previously stated selfish objectives. Note that all members of a specific cross-organisational group must buy into and support the group's statements of joint objectives and the group must not discount any one organisation's selfish objectives without clear and full agreement. This session will require much discussion, fusing (say) 25 selfish objectives into 5 statements of joint objectives. The discussion is an important part of the development of cross-organisational understanding. Note that this is not an exercise in compromise. No organisation should be asked to compromise on their objectives but should align their objectives with those of other team members in pursuit of a common goal.

At the end of this cross-organisational session, the facilitator will reconvene the team and ask each cross-organisational group to display and feed back their joint objectives to the full workshop team, again taking around five minutes each. The facilitator will encourage questions, comment and discussion on the objectives and should also check that all original selfish objectives have been taken into account, either explicitly or tacitly within the changed wording of the joint objectives.

At this stage, assuming four cross-organisational groups, there will be four sets of five joint objectives, a total of twenty statements of joint objectives. These may be similar but it is likely that each group will have made their statements in slightly different ways.

The third and final session is held with the full team and is focused on assisting them to turn their statements of joint objectives into a partnering charter of mutual objectives (see Fig. 16.1). This may take up to an hour and is an important stage in obtaining the team's buy-in to the specific wording of the charter. The facilitator will lead the team to develop the wording of each individual phrase, bearing in

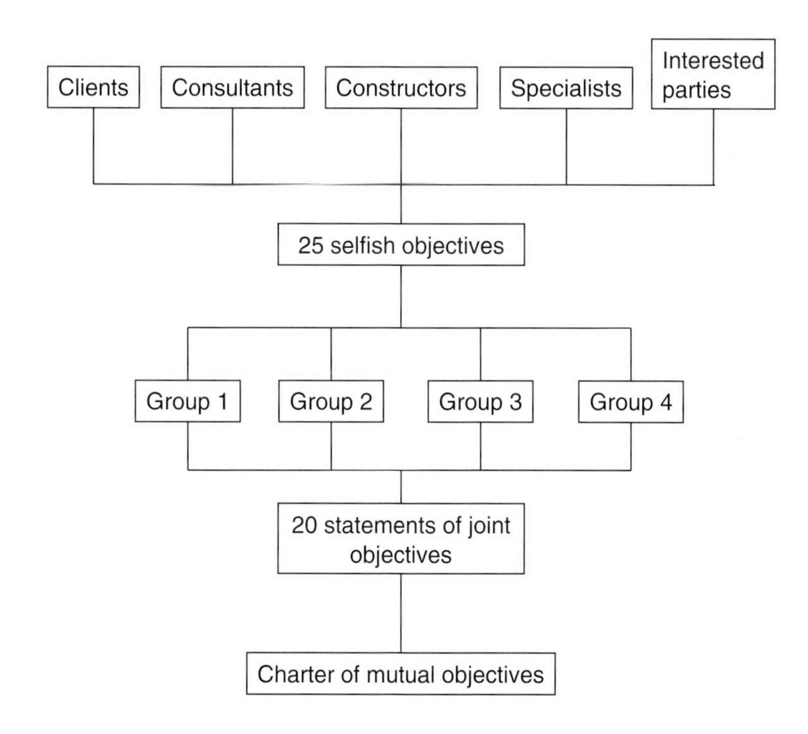

Figure 16.1 Developing mutual objectives.

mind that the charter may be displayed on site, in offices and in publicity. The format and language should therefore be meaningful and relevant to all who may see it.

We use a range of media to support the development of the charter in full session, depending on the facilities available at the venue. Each has its advantages and disadvantages:

- A flip chart enables the facilitator to make changes to the wording whilst retaining the original but after a number of changes the flip chart will have many crossings-out and it may be necessary to rewrite in order to clarify the agreed statements.
- A white board enables some semblance of tidiness to be retained but once redundant wording has been erased it is lost (unless a second facilitator has been taking notes).
- A projected computer is faster and more legible than handwritten flip charts and does not require the completed charter to be re-typed on completion but, as with the white board, once the wording has been deleted, it may be lost.

Once the charter wording is complete, the team will move onto the next workshop stage (normally issue resolution). The facilitators will tidy up the format (but not the wording) of the charter during the next break and print out copies for the team to review. Typically, we will review the charter with the team after lunch and, subject to team agreement, print out a fair copy which all team members will sign at that time, demonstrating their commitment to the team's mutual objectives. Leaving the signing until the end of the workshop runs the risk that some members may be called away and may lose the opportunity to sign.

We have already mentioned that senior members of the partners need to be present in order to present their organisations' objectives, lead their group and commit to the charter. Taking a draft partnering charter away from the workshop for review and signing (or amendment) by senior management is demeaning, deflating and a major disincentive to partnering and integrated teamworking.

As an example of the development of a partnering charter of mutual objectives, we have extracted the following from the report of the initial partnering workshop for the Open University New Library (January 2002).

In the first session, the team was divided into two organisational groups to define their six selfish objectives for this project. These were fed back in full session as follows:

Consortium Group (Swanke Hayden Connell; Davis Langdon; Buro Happold; Galliford Try)

- ❑ profit and efficiency
- ❑ a high business profile from an exemplar project
- ❑ a satisfied and happy client
- ❑ transferring understanding and quality from design to completion
- ❑ lower defensive barriers without risk
- ❑ enjoyment – satisfaction without stress.

Client Group (OU Estates; OU Library; Malcolm Reading and Associates)

- ❑ be within budget
- ❑ completion on time
- ❑ meeting the brief completely
- ❑ achieving best value
- ❑ achieving a defect-free smooth occupation
- ❑ achieve a good learning experience.

The team reviewed each others' objectives and were then split into two cross-organisation groups to determine expressions of joint objectives derived from the selfish objectives:

Group 1

In the design construction and occupational life of the building we aim to achieve:

- ❑ best value by meeting the brief completely and transferring understanding to achieve ultimate quality
- ❑ high business profile from a good learning experience and an exemplar project
- ❑ real completion on time within budget and achieving profitability
- ❑ satisfaction and enjoyment for all
- ❑ honesty and reduction in defensive barriers.

Group 2

Achieving the brief requirements through:

- ❑ a seamless and efficient design, approval and construction process
- ❑ collective and timely problem resolution

- ❑ enjoyment, satisfaction and pride
- ❑ successful and respected business approach

...achieving defect free, smooth occupation and operation.

Finally, in full session, the team fused the joint objectives into a partnering charter which all delegates signed (see Fig. 16.2).

Open University New Library Partnering Charter

In the design, construction and life of the building we will support each other to achieve:

- Best Value whilst meeting the brief completely
- Sharing understanding to achieve efficiency and quality
- Collective and timely problem resolution through openness and the removal of defensive barriers
- Enjoyment, satisfaction and pride in an exemplar project
- High business profile from a good learning experience
- A successful and respected business approach
- Defect free, smooth occupation and operation

Figure 16.2 Typical partnering charter.

Through the process of developing a partnering charter of mutual objectives, all the team members are actively involved, identify their objectives, understand the objectives of their partners and discuss how to align these so that the whole integrated team is pulling in the same direction and maximising their efforts for the benefit of the project.

17 Issue Resolution

There will always be problems on construction projects. *Trusting the team* (Bennett & Jayes, 1995) identified that the second essential feature of successful partnering relationships was an agreed method for resolving problems. These problems or issues may be driven by a number of factors which may include onerous contract conditions, personality clashes or factors outside the team's immediate control.

To maximise efficiency in the team, it is critical for the integrated team members to establish a clear and robust process at an early stage in the relationship in order to identify and resolve issues clearly and swiftly, before they develop into major problems. It is far more effective to resolve issues promptly where and when they arise rather than escalate them to higher authority or allow them to fester.

A clear and robust issue resolution process that is set up by and understood by all team members at the initial partnering workshop will aid communication as it will define and clarify roles, responsibilities and authority.

In the unfortunate event that an attempt at resolution fails, it is necessary for the team members to understand and adhere to an agreed process for escalation to the next level of authority, simultaneously in all organisations. A measure of success of the issue resolution process is how few issues reach the top level in the issue resolution ladder.

During the initial partnering workshop, the team will create a matrix that will identify a limited number of levels of empowerment for decision making and develop a process that will enable all team members to resolve issues or to escalate these in an appropriate manner to higher levels of authority. The result of this stage of the workshop will be to ensure that all team members understand the following issue resolution process:

- team members who identify an issue should resolve this within an agreed timeframe with team colleagues at the same level on the issue resolution ladder
- if such resolution is not possible within the agreed timeframe, those who cannot resolve the issue will jointly raise it to those team members named on the next level of the ladder.

At the start of the issue resolution stage, the facilitator will divide the team into mixed-organisation groups, each comprising no more than seven members, and give each group a pre-printed matrix as shown in Fig. 17.1.

Each group will be tasked with producing, within 15 minutes, a completed matrix filled with the names and roles of those team members empowered to resolve issues at specific levels.

After the 15 minute group stage, the team will reassemble and the facilitator will take feedback from each group. Using a flip chart or projected computer, the facilitator will complete the matrix in full view of all the team. At this stage it is advisable to use a different colour for each group's feedback. Note that there may be discrepancies between each group's understanding of the roles of different members of different organisations. After feedback from the groups, the facilitator will encourage discussion on the different opinions and attempt to reach a consensus across the full team.

Our experience is that the first draft of the issue resolution ladder is completed within 30 minutes and the team is ready to move on to the next stage. However, the facilitator should ensure that the team works through two or three specific scenarios from site/project level through to core group to simulate and test the process and to confirm full team understanding. In carrying out these exercises, the team will ensure that those on the ladder are not only empowered to resolve issues at their level but are willing to do so and are actively encouraged by management and the core group. We have frequently found that this is a rich opportunity to clarify confusion over roles and responsibilities within the team. We have had occasions, for example, when it has been revealed that directors of constructors and clients are being relied on to take site decisions. This drives delay in the resolution of issues with knock-on effects on efficiency and profitability.

	Client	Constructor	Consultants	Other
Core Group				

If agreement is not reached within ___days/weeks then the issue is jointly referred to the Core Group.

Management				

If agreement is not reached within ___days/weeks then the issue is jointly referred to Management.

Project				

Figure 17.1 Issue resolution ladder.

Once the issue resolution process is set up, it should be communicated to all partnering and integrated team members by the core group. They should review the successes of the issue resolution process on a regular basis noting that the more issues that can be resolved within agreed timeframes at the level at which they arise, the more effective the integrated team will be.

The team members at the initial partnering workshop may find that it is not possible to reach a consensus on the issue resolution process within the allocated stage. The facilitator may choose to extend the stage in order to sort this out as any delay to agreement of the process will prejudice the speedy resolution of any issues that arise. In the event that the team cannot reach consensus, the facilitator should ensure that a clear action is placed on the core group to resolve the matrix and the process outside the workshop within a specific time (say two weeks) and publish it to all team members.

Partnering teams need good communications and early warning mechanisms that alert the whole team to potential and actual issues. During the early days of partnering in the UK (1995 through to 1998), many best practice books developed their own diagrams for the issue resolution process. We have developed these to include the further stages of communicating the issue and solution (and changing processes where appropriate) in order to avoid repeat occurrences of the same issue (see Fig. 17.2).

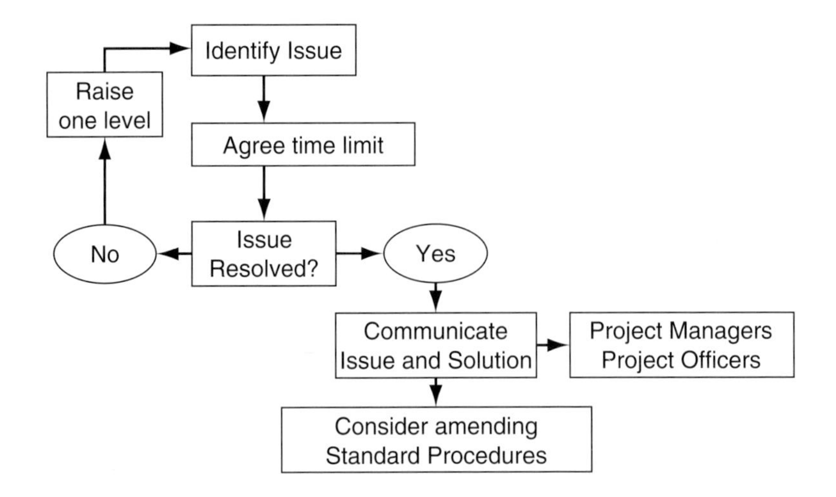

Figure 17.2 Issue resolution process.

In order for the team members to resolve issues early, it is important for them to identify (and then to resolve) the real problem rather than the symptom. Time spent here will focus the team's attention on addressing the root causes of issues rather than applying *sticking plaster* solutions to symptoms.

We developed an early warning STEP-sheet (State, Trace, Evaluate, Propose) to standardise the presentation of issues. We suggest that those team members who have identified the issue should jointly complete this STEP-sheet and seek to resolve the issue within their own set time limit:

- ❑ Stating the issue
- ❑ Tracing the cause (why are we in this situation?)
- ❑ Evaluating the potential impact and likelihood (what's going to happen if we don't sort this out and how likely is it to happen?)
- ❑ Proposing solutions, stating potential costs and benefits (how can we get out of this situation and what costs and benefits will there be?) and identifying agreement reached.

The act of jointly setting out the issue on this sheet may help the parties to solve it without escalation to higher authority (see Fig. 17.3).

Issue identified by: Date: Final date to resolve the issue:
• State the issue
• Trace the cause
• Evaluate the potential impact and likelihood
• Propose solutions stating potential costs and benefits identifying agreement reached

Figure 17.3 STEP sheet.

If the conclusion is agreement on the part of the respective parties, they should communicate this resolution to the core group who can disseminate the solution to others as appropriate, transferring knowledge to all organisations.

If the conclusion is not agreement, this sheet should be copied to those identified on the next level up in the issue resolution ladder with a request to resolve the issue by a specific date.

If decisions prove, in hindsight, to be ill-judged, it is incumbent upon senior management to review the issue and the solution with those who made the decision, not attributing blame but ensuring learning for future projects and the individual concerned. This approach is likely, in the long term, to prove less costly than referring all decisions to senior management (who are, in any case, unlikely to be in possession of all the facts at the time the decision is required). This culture of empowerment and no-blame will encourage trust from within the team and lead to more issues being resolved at the most appropriate level.

When escalating issues, team members should ensure that alternative solutions are clearly set out for decision at the next level, including the costs and benefits of each. It is our opinion that, for the long term benefit of the team, if a referral does not include solutions, management should reject it and insist that those referring the issue re-present it with alternative solutions. Whilst this rejection may cause a delay on the first such occasion, we believe that it encourages team members to adhere to the process (proposing solutions) or, more effectively, to resolve the issue themselves where it arises, speeding issue resolution and developing a solution-focused culture.

18 Partnering Champions and the Core Group

It is critical to the success of the partnering relationship that each organisation has a partnering champion who is fully committed to partnering and is empowered to take responsibility for understanding and implementing partnering within and beyond the organisation. Partnering champions will be senior members of the organisation, willing and able to challenge established practices. They must have the explicit support of senior management and the respect and support of their peers. These partnering champions are likely to form the core group which will steer the integrated team to deliver better value for the project and all organisations.

The requirement in PPC2000 (The Association of Consultant Architects Ltd & Trowers & Hamlins, 2000) for all parties to, '...work together and individually in the spirit of trust, fairness and mutual cooperation...' is an aspect that may require a major cultural shift on the part of organisations and team members using the contract for the first time. It will fall on the shoulders of the champions to initiate, encourage and maintain this cultural shift.

The partnering champion should be appointed by senior management as soon as an organisation commits to placing any work through the partnering route. Senior management should select a partnering champion who is enthusiastic and committed to the role, being prepared to commit their time and energy. It is important that champions understand and are able to convince their colleagues of the value criteria of their business if they are to overcome the lowest tender price culture that still exists in many public and private sector organisations.

In order to convince others of the value of a cooperative teamworking approach to construction, champions must have the ability to communicate and make presentations. They will have to explain the partnering process to others at all levels in the organisation, listening to and addressing concerns, whilst keeping up to date with the latest partnering practices and disseminating information, especially the early successes. Champions will play leading roles in partnering workshops and team building exercises.

Whilst the champions will have to commit considerable personal energy into their role, the organisation must provide moral support and resources. Organisations must empower their champions to take decisions, place orders for the facilitator and adviser, book workshop venues and follow through changes to the structure and processes of the organisation and of the partnering and integrated teamworking arrangement. Such empowerment will add credibility and build trust in the champions and help to ensure that future initiatives are supported by the rest of the team.

Champions should be able to persuade their colleagues as they should not be in a position to dictate in an empowered structure. They should not impose their ideas or the ideas of senior management but facilitate the process of culture change by encouraging discussion and obtaining team buy-in to decisions.

At the initial partnering workshop, each organisation should identify the name and role of their partnering champion. It may be beneficial for these names to have been shared and discussed between the organisations before this stage in order to choose not only committed individuals but champions who can work well together.

After the formation of an integrated team, the partnering champions of the various partnering organisations will work together and form the core group who will oversee and steer the team to deliver their strategic goals. PPC2000 defines the function of the core group as to, '... meet regularly to review and stimulate the progress of the project and the implementation of the partnering contract.' The same principle applies to the role of a core group in a framework or a term contract – arranging regular meetings, stimulating the team and ensuring implementation of partnering initiatives.

The core group is assembled at the start of the contract and PPC2000 makes it the responsibility of the client representative to call, organise, attend and minute regular meetings of the core group.

The contract further states that organisations must ensure that members attend core group meetings and allow core group members time to fulfil their agreed functions which will include following through agreed actions (see Fig. 18.1).

The core group members will probably form the top level of the integrated team's issue resolution ladder and should address only those issues that are escalated to them from the lower levels on the ladder. The core group should keep a register of any issues raised to their level to ensure that the issue resolution process is working. If the core group cannot reach consensus on issues, the point will have been reached at which issues must be referred outside the partnering team to the partnering adviser or to mediation. This is a major step and one that will not be taken lightly as it will signal the failure of the integrated team to resolve its own issues.

To pre-empt the situation in which issues have to be referred beyond the team's agreed issue resolution ladder, the core group must make every effort to keep their fingers on the pulse of the relationship. Through support for team members in resolving issues, the core group will develop a solution focused culture in the team.

As the core group is set up early in the relationship, it will be responsible for ensuring that the whole team know and understand the purpose of the selected key performance indicators (KPIs). If the KPIs are capable of being measured regularly through the contract, the core group will be able to identify when there are peaks or troughs in the relationship and apply corrective action through incentives or rewards. Note that this action applies not only to the progress of the contract (under or over budget or time) but also to relationships, for example, assessing whether or not trust is growing.

A straightforward way to monitor trust in the team is for the core group to send out a questionnaire on a monthly basis asking the team members to respond to the following question: 'On a scale of 0 (unacceptable) to 10 (excellent) how do you feel the partnering team members are doing at working together and individually in the spirit of trust, fairness and mutual cooperation for the benefit of the project?'

A simple spreadsheet analysis would show high, average and low scores and the trend since the last review – raising the core group's awareness of the state of the relationship and prompting them to take steps to maintain or improve the current situation.

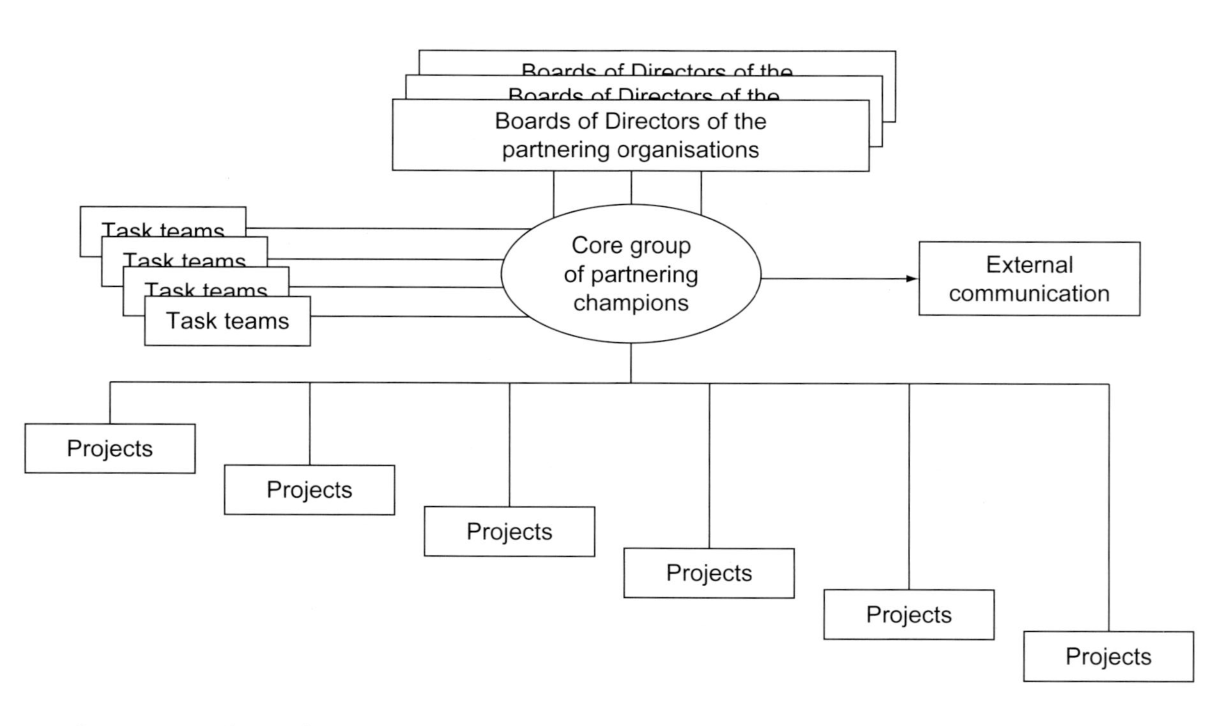

Figure 18.1 Core group relationships.

We have been fortunate to have worked with many organisations whose partnering champions have been extremely effective. The success of the partnering approach to Shropshire County Council's Highways Maintenance contract with JDM Accord owes much to the proactive drive of the two partnering champions – Chris Edwards (SCC) and Tim Bebb (JDMA) who were appointed very early after the initial partnering workshop. Within four months Chris and Tim had arranged partnering awareness training and continuous improvement sessions for all management and staff of both organisations (including the administration teams and tradespeople) and set dates for future continuous improvement workshops.

During the initial training and continuous improvement workshops the delegates identified more than 60 potential opportunities for improving the service including cutting out waste, clarifying tasks, improving communications through developing informal and face-to-face contact between team members. The champions visited team members at their places of work, updating them on progress and picking up further successes and ideas for improvement. They also published newsletters to the team highlighting successes and acknowledging the contribution of individual members within the team. Further information on the Shropshire Champions is set out on the Partnership Sourcing Ltd website www.pslcbi.com.

Delivery of better value through partnering and integrated teamworking demands good interaction between all team members. Whilst the relationship needs the commitment of each and every team member, the choice of partnering champion or core group member is, in our experience, critical to the success of the arrangement as it is the champions or core group who drive forward the culture and practices of partnering and integrated teamworking.

19 Continuous Improvement

If you do what you always did you'll get the results you always got. Continuous improvement is the ongoing pursuit of added value through reduction of waste and the provision of greater satisfaction of user needs. It is the feature that differentiates a full partnering relationship from a cooperative integrated teamworking culture and is the third essential feature of successful partnering relationships identified in *Trusting the team* (Bennett & Jayes, 1995).

Adding value is the reason for changing processes and cultures from a low purchase price tactical approach to a best value strategy. However, without sustained effort instigating, implementing and measuring continuous improvement, the integrated team is in danger of succumbing to the challenge that cheaper performance could be achieved through competitive tendering on price. In order to stay together and maintain the opportunity to improve, the team must identify success, clearly measure the improvements they have made and convert these to a value base that is understood by all, especially those who would prefer the traditional route. For the integrated team to build on their success, the core group must communicate the value of these improvements internally to the whole team and externally to all interested parties.

As a first step in continuous improvement, the final session of the initial partnering workshop should be devoted to identifying key opportunity areas and, where time permits, developing these opportunities. Tools for developing continuous improvement opportunities include value management, risk management, lean thinking and cross organisational learning, all of which are developed later in this book.

In long-term relationships (for example, term contracts, frameworks or projects that will last more than six months) the full integrated team should hold regular continuous improvement workshops, away from the day to day pressure of the contract, to review successes, identify opportunities and drive continuous improvement.

The team should seize every opportunity in the early days of the relationship to identify and capitalise on the easy wins to prove the value of the integrated team approach. They should prioritise the *big win, little effort* opportunities over the *big win, big effort* ones that can be addressed in future continuous improvement reviews.

Many organisations and individuals find it difficult to quantify the added value of partnering and integrated teamworking except by identifying the hard savings such as bricks 10% cheaper or the same building costing 5% less than the previous year. However, a substantial proportion of the added value from partnering and integrated teamworking is derived from the reduction of duplication and waste of resources and from other efficiencies, such as whole-life cost and sustainability opportunities identified by team members. Within a drive for better value there should be '. . . a shift in emphasis from initial purchase costs and short-term savings to the examination of whole-life costs and establishing longer-term objectives to ensure overall best value' (Chartered Institute of Public Finance and Accountancy, 2003).

It is a fact that most projects are justified on financial criteria. Despite the push towards quantification of *soft* or longer-term benefits, these may fail to attract the attention of auditors or financiers unless the benefit can be expressed in a monetary form. In order to aid the efforts of teams dedicated to driving better value, we have set out below some hard values of soft benefits for consideration, discussion and development by the team. We have assumed £320 per day (£40 per hour) for the total cost of management and £100 per day (£12.50 per hour) for the total cost of clerical and administration staff working on the basis of a 200-day working year and including all overheads and office space.

The team may wish to change the rates used but should ensure that the relatively conservative timings used in the examples are also reconsidered. The purpose of these examples is to demonstrate that the team needs to look at the total costs of purchase and to think outside the constraining box of tendered rates if the team is to drive better value for all.

- Fewer letters – each typed letter takes 15 minutes management time drafting and checking and 20 minutes clerical time typing, correcting, filing and posting = £14.16 per letter plus the cost of file storage, paper, postage, etc., say £15 per letter.
- Fewer meetings – each meeting lasting three hours with ten management costs 30 hours at £40 = £1200 per meeting plus the cost of meeting rooms, coffee, minutes, etc.
- Minutes of meetings – each set of minutes takes two hours management time in drafting and checking and an hour clerical time typing, correcting, filing and posting = £92.50 per set of minutes.
- Tenant complaints – Each complaint has to be taken and logged, passed with details to the appropriate manager, researched, answered and copied to appropriate parties (elected members, file, etc.). Say two management hours and one admin hour per complaint plus two letters (as above) = £92.50 dealing with the issue plus two letters at £15 each = £122.50 per complaint.
- Defects – each defect costs a tenant complaint (£122.50) plus the cost of sending out a surveyor or manager to check and re-order a repair (one hour = £40) plus the cost of processing an order (say £15) plus the cost of the specialist's visit = £177.50 per defect plus £30 for the visit of the specialist = £207.50 per defect.

These are only examples and don't take into account all issues. The £1200 meeting, for example, assumes that the meeting is managed effectively and doesn't take into account the further potential waste caused in teams that meet but don't discuss, those that discuss but don't decide or those that decide but don't do.

The cost of defects can only be passed on through the supply team to the eventual client or end-user. It is in the interest of all members of the integrated team to eliminate these costs. In a recent post-project review, the site agent from Mansell identified that the effort of planning and keeping to a clear sequencing of trades when completing a multiple bedroom/bathroom facility ensured that revisits were eliminated, saving considerable time and resource costs.

Quantifying the benefit of early completion in the private sector, it can be calculated that predictable early completion of a capital project for a 40-bed hotel = 40 rooms @ £100 × 7 nights × 80% occupancy = £22 400 per week plus the opportunity costs of food and beverage sales. In the public sector, predictable early completion of a capital project for a rented social housing block of 20 units would gain lets of 20 units @ £60 per week per unit = £1200 per week.

In reducing the resource waste associated with slow issue resolution, members of integrated teams should understand that preparing the facts, outlining the case and referring an issue for decision to a higher level on the issue resolution ladder, can take in the order of two hours per referral. At £40 per hour, this costs the team £80 per referral per team member involved. A swift joint decision by two empowered team members, avoiding referral, can thus save £160 of resource time.

Some estimates put the labour waste associated with standing time on site as high as 30%. This would include waiting for instructions, deliveries of materials and a break in the weather. If, on a project of £1million, the site labour content is 40% of the total (£400k), this waste is worth £120k. It must, therefore, be worth an integrated team spending two days in a workshop to halve this waste.

Material waste is all around on construction sites. It has been identified (Construction Confederation, 1999) that,

> 'A breakdown of skip contents has shown that the materials that give rise to the greatest amounts of waste are those which are the easiest to recycle (hard materials and timber). Up to 25% of waste produced on construction sites could be minimised relatively easily, increasing profits by up to 2%.'

Increases in landfill tax have probably made the 1999 figure of 2% a substantial understatement.

Individual partnering and integrated teams should be encouraged to identify the cost of waste in whole-life costs that can be addressed within their project. The team could consider, for example, an environmental sustainability proposal in an office development. If such a proposal were to reduce sickness absence in an office by just two management days a year, measured against previous records for all staff, this would benefit the company by two days @ £320 = £640 per annum. Over five years (not discounted) that would equate to a £3200 added value benefit. So continuous improvement in sustainability can also be quantified and added to a business case.

The opportunity for continuous improvement in construction projects is vast. In addition to the more obvious waste of materials, we have shown that there is waste in correspondence, meetings, complaints, defects and on-site labour. Teams will have other examples which may include unstreamlined processes and multiple handling.

Through joint working and planning and a proactive drive for continuous improvement, the integrated team can identify opportunities to eliminate the activities that do not add value and focus on those that do.

20 Benchmarking and Key Performance Indicators

Without clear measurement of performance against key value criteria it is difficult for any team to determine how successful they are. Traditionally, the industry has tended to judge construction contracts by determining, for example, whether the final account was within budget or whether the project was delivered on time. There has also been a tendency to use past project information to blame one or other of the participants rather than to take on team responsibility for the project.

Benchmarking is defined at www.constructingexcellence.org.uk as, 'a method of improving performance in a systematic and logical way by measuring and comparing your performance against others and using lessons learned from the best to make targeted improvements.'

Key performance indicators (KPIs) are defined at www.constructingexcellence.org.uk as, 'the measure of performance of an activity that is critical to the success of an organisation' and will probably be based on the value criteria of the organisation.

The integrated team should benchmark its performance by taking the results of its key performance indicators and comparing these with the performance of other organisations or with its own past performance as appropriate. Constructing Excellence publishes graphs of industry performance against each of the industry standard KPIs in order that teams may assess their performance against the best.

In order to drive continuous improvement, it is important for the team to identify targets that are achievable. We suggest using use *best-in-class* as a benchmark. Knowing that the best-in-class performance has been achieved by other teams helps to counter the argument that it can't be done.

In an increasingly competitive industry in which selection is increasingly based on criteria other than (or in addition to) price, a structured KPI process will benefit the integrated team. The KPI process will enable the team to demonstrate better value by analysing and comparing its project performance against other teams and projects on a range of key criteria, helping the team to identify and address successes and opportunities. Clear, objective measurement of past performance can thus be used to ensure rational selection of the better performing teams, address areas of weakness and improve future project performance.

The three letter acronym, KPI, has become such a standard term that it is important to define the meaning and use of KPIs in an integrated team environment. KPIs are indicators of performance in key areas of the project, programme or partnering arrangement as defined and agreed by the partners. In the earlier chapters we identified that the number of key value criteria for clients and suppliers would probably be no more than ten for any relationship. We would suggest that the number of KPIs is also no more than this number as the benefits of measuring and monitoring an increased number may be outweighed by the costs incurred.

If an integrated team wants to compare its performance with the rest of the construction industry, the core group should consider benchmarking themselves against some of the industry standard KPIs available at www.kpizone.com. The formulae are already set out and Constructing Excellence publishes annual charts of industry performance against which the team can compare their own performance.

The initial set of construction industry KPIs was based on the targets for continuous improvement set by Sir John Egan's Task Force in *Rethinking construction* (Egan, 1998). These KPIs are generic and can be used for almost all projects of whatever size or complexity. The Constructing Excellence KPI handbook includes standard formulae for calculation of each KPI so that the performance of different projects can be compared across different sectors. The ten standard all-construction key performance indicators are:

1. client satisfaction with product
2. client satisfaction with service
3. defects
4. predictability – cost

5. predictability – time
6. construction time
7. construction cost
8. contractor profitability
9. contractor productivity
10. safety.

In addition to the all-construction KPIs there are sets for:

- housing
- respect for people
- environment
- construction consultants
- M&E contractors
- construction products industry.

In the event that the integrated team has a programme of work, they may decide to set their own KPIs in order to compare their projects in the current year with those of previous years. They may choose to develop sector-specific or other indicators that are not included in the industry standard packs. In this case, the core group should identify the KPIs early. Some draft KPIs may be included in the partnering documentation sent to bidders and some may be elicited from the bidders in the selection process.

There is no benefit in leaving the selection of KPIs until late in the team relationship. If the measures are key to the organisations, the project teams need to know what they will be judged against and given the opportunity to perform well. If the client has questionnaires that support a KPI (for example, customer satisfaction), they should be shared with the integrated team at the start of the project. We make a point of sharing such questionnaires with integrated teams at the initial partnering workshop. In this way, the team can focus on the key areas of the project on which they will be judged and which reflect the client's key value criteria.

At a meeting of the core group, in the initial partnering workshop or in a separate KPI workshop, it is important that the integrated team agrees a limited number of KPIs, defines the formulae to be used, identifies when each KPI will be measured (e.g. monthly, annually, at project completion) and agrees who will carry out the data collection and analysis. Within the KPI workshop, the team should set up a robust method of communicating the KPIs to all the team before the start of any project and a process for sharing the results as they are collected so that trends can be identified early.

Many partnering arrangements have KPIs that are set up solely to measure the performance of (say) the constructor or other suppliers. This is perceived as unfair by many organisations and, for successful partnering and integrated teamworking, the KPIs should be indicators of performance of the whole team.

In developing the KPIs, the team should follow these steps:

- review the function of KPIs
- elicit KPIs from team members, listing these on a flip chart
- encourage discussion
- prioritise no more than (say) ten KPIs
- identify which KPIs are industry-standard and for which the team can, therefore, utilise the industry standard formulae.

For any non-standard KPIs the facilitator should:

- break the team into groups and assign one KPI to each group for development
- task each group with developing the formula for calculating their KPI, identify what data is required and how often this must be collected, by whom and when
- ask each group to feed back the information on the KPI they have developed
- encourage discussion and seek clear and concise agreement on each KPI
- trial run the calculation with dummy data.

If it proves difficult to agree the measurement process, the team should consider whether the KPI is appropriate or relevant. The team should also note that a non-standard KPI that requires complex data collection or calculation may be counter-productive and demoralising as the team may spend a disproportionate amount of time measuring performance.

At the initial project workshop, the project manager should review the rationale for the KPIs, the formulae and the measurement process with the integrated team.

At each project site meeting, KPIs should be an agenda item and any issue that may impact on the scores should be raised, discussed and resolved.

At continuous improvement reviews in frameworks and term contracts, the full integrated team should receive a short report from each project completed over the previous period. This report should review each project's successes and opportunities for improvement and share the radar chart of KPIs. This direct communication across the integrated team helps to share knowledge and build trust.

In addition to the end of project KPIs identified above, the team may want to measure performance on an ongoing basis throughout the project. We use the statements of mutual objectives set out in the partnering charter to measure the team members' perceptions of their performance at continuous improvement workshops.

When issuing the agenda for the post-project review, the project manager should remind the project team to bring all relevant data to the review. At the post-project review the project manager should lead the project team in compiling the KPI data on a standard form and in developing a radar chart (see Fig. 20.1). The project manager should send the completed KPI radar chart, together with the report on the post project review, to the core group.

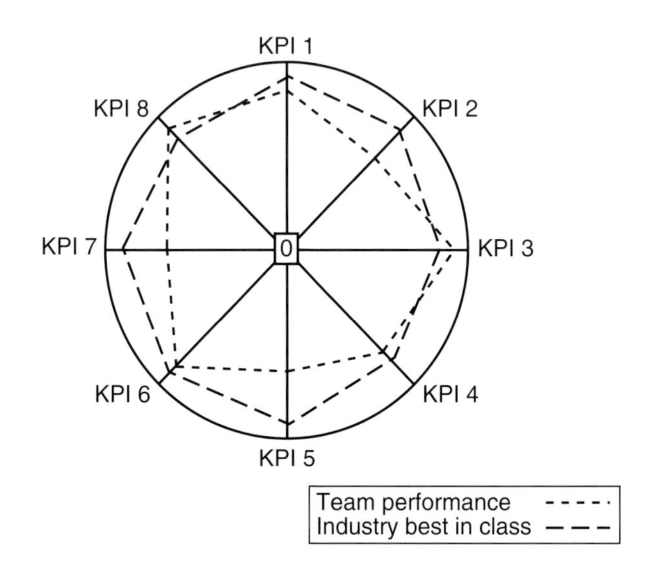

Figure 20.1 Typical radar chart of KPIs.

Note that the conventional presentation is to place the lowest score in the middle of the chart and the highest scores on the outer rim. It can be seen at a glance in the chart above that the team has outperformed the industry best-in-class in KPIs 3 and 8 whilst significantly underperforming in KPIs 2, 5 and 7.

Whilst we use radar charts to enable us to assess project performance at a glance, it is important to view the KPIs in context. For example, a low score on timeliness may reflect a team decision to complete late in order to achieve zero defects. Equally, a low score on cost predictability may reflect a team decision to spend additional funds to accelerate a programme. The report accompanying the radar chart should include the story behind the KPI scores, especially where these are significantly high or low. In this way the core group will identify where the team has succeeded and where there are opportunities for continuous improvement. Whilst it is important to improve in underperforming KPIs, the team must continue to perform well in the successful KPIs. The target must be to exceed the industry best-in-class performance in all KPIs.

Comparison with industry performance helps integrated teams to assess how well they are performing in key areas and may help to bring sceptics on board. Where appropriate, it enables teams to prove that their performance delivers better value than (for example) a competitor who delivers late and with high levels of defects and customer dissatisfaction – even though they are cheap. Also, the fact that the team has the confidence to measure its performance and has a willingness to improve should set it apart from those teams that do not measure and cannot prove how well they perform.

21 A Programme of Partnering and Integrated Team Workshops

In order to achieve measured and measurable continuous improvement, it is necessary for partnering and integrated teams to meet on a regular basis. The core group should establish a structured programme of workshops and team events for the duration of the relationship. This will assist the team in growing cross-organisational understanding, building on previous successes and developing opportunities for further improvement to add value for all team members. Regular face-to-face contact also helps to break down barriers and to reduce misunderstandings.

The programme of workshops and team events should be set up at the first opportunity, once the decision has been taken to proceed with a partnering and integrated teamworking arrangement. The responsibility for setting up the programme will normally rest with the client or the client's representative. The partnering contract PPC 2000 (Association of Consultant Architects Ltd & Trowers & Hamlins, 2000) sets a responsibility on the client representative to organise value and risk management exercises, partnering workshops and a post project review.

There is a wide range of facilitators who can provide appropriate support for specific workshops. Their costs and facilitation styles will vary and availability for a specific facilitator may be at a premium, requiring pre-planning by all involved. We have set out guidance on selecting facilitators in an earlier chapter.

The team should be aware that there is a cost to the workshops. In addition to the hire costs of a venue, the core group must consider budgeting for specialist facilitation. Hired venues and outsourced facilitators will be obvious costs against the budget of the project or individual organisations. In-house provision of facilities and facilitators will, typically, incur opportunity costs that may not be charged to the project. However, when the costs of workshops are measured and compared with the added value of an integrated team working in harmony (or, conversely, the real and opportunity costs of a team not working in harmony), the business case will almost invariably justify setting up and maintaining a structured programme of team workshops.

In addition to outgoing fees, there is the resource cost of such a programme. For an £8million project (six months in design and fifteen months on site) there will be in the order of ten workshops, each with around fifteen attendees. In this case, the resource cost is 150 management days. Whilst this may seem excessive at first sight, the regular structured workshops should lead to a reduction in the number of ad hoc meetings between the various individual members of the team and, through full-team attendance, ensure greater cross-organisational understanding. The reduction in ad hoc meetings should be quantified by the core group and presented as a benefit of the integrated team working and partnering approach to set against the costs of workshops.

Many teams may look at the apparent cost of workshops with horror and decide to save costs by meeting once in a while or by not planning future meetings. This is a mistake. The cost of ten workshops for the project outlined above (initial workshop, value and risk management, six quarterly continuous improvement workshops and one post-project review) may be in the order of £20k for facilitation and £5k for venue hire. However, bearing in mind that this project may be budgeted at around £8million, it only needs just over 0.3% value enhancement from working together to repay these costs.

In addition to structured project-focused workshops, the core group should consider the need for the team members to develop a common cross-organisational culture of partnering and integrated teamworking. This may require a training programme to include not only the technical members of the team but also their management, directors and any other interested parties who may impact on the successful delivery of the project.

It has been identified (Bennett & Jayes, 1995) that the total costs of partnering, including training and resource time, amount to approximately 1% of the project cost, although this will clearly depend on the scale of the project and on the number and frequency of workshops. The resulting benefits from partnering and integrated teamworking have also been measured. These are in the range of 3–10% for project partnering and 10–30% for longer term, strategic relationships (Bennett & Jayes, 1995). The return on investment is therefore substantial.

The programme of workshops should be structured with the intention of benefiting the delivery of the project. These are not client workshops or design team workshops – they are integrated teamworking workshops for all team members.

The team members should be informed of the programme and the specific dates of workshops as soon as possible. They should be actively encouraged by their managers and directors to reserve these dates in their diaries and to make every effort to attend in order to maintain the drive towards an integrated team. It is important that all members attend the workshops that are arranged rather than send a representative as continuity of personnel is paramount when building relationships between organisations.

Having drawn up a programme of integrated team workshops, it is essential to adhere to it if optimum results and team motivation are to be achieved. Cancelling workshops (whether as a result of cost or for any other reason) sends a message to the team that partnering and integrated teamworking is a low priority. Cancellation of workshops dampens enthusiasm, loses the team's momentum and risks the relationship.

Social events should be an integral part of the workshop programme, as partnering and integrated teamworking depends on the interaction of team members – getting to know one another, how they work and how they respond to challenges.

Once a partnering and integrated teamworking route is chosen, the team should be brought together as early as possible at the initial partnering workshop in order that all share a common understanding of the client's and each other's value criteria and the ways in which these are to be delivered. At this early stage, all team members should be allowed the opportunity to input their own views and suggestions on delivering better value.

Following the initial partnering workshop, the team should reconvene in value and risk management and continuous improvement workshops. These will all assist in building the partnering team ethos as well as defining and refining the scheme. Following these workshops, it may be appropriate to set up specialist task groups to address further specific topics. The proposals and results from these task groups should be fed back to the team through the core group/ partnering champions.

We have set out below (with the permission of Andy Ward at JDM Accord) a good example of the work of a task group following a continuous improvement workshop proposal. In addition to encouraging the implementation of the initiative, the communication was used to clearly set out the improvement process and identify the outcome and benefit to the organisations.

JDM Accord – Shropshire CC Partnering Initiative – Office Paper Recycling

We have started to recycle unusable office paper at the Longden Road depot. SITA have provided twelve cardboard recycling boxes that are located throughout the offices, a supply of plastic liners to hold the waste paper, a skip specifically for waste paper located outside the workshops and a disposal point at Granville where the paper will be bailed, compacted and forwarded to recycled paper manufacturers. The message is first to re-use paper if at all possible and second to recycle it by placing it in the cardboard bin. You will be surprised by the amount we discard. By participating in this recycling activity we are all saving JDM Accord the cost of disposal and we are helping in diverting biodegradable material from landfill.

Following implementation of the scheme, the office cleaner reported a 60% reduction in the general office waste, indicating the amount of paper discarded.

During the remainder of the project, the team should meet on a quarterly basis in continuous improvement workshops which may be targeted at specific areas of the project. These may also be an opportunity to develop the team through non-project teambuilding exercises.

After handover, the team should meet again for a post-project review to celebrate success, close out any remaining issues, agree and report on KPIs and take forward the successes and opportunities to their next projects.

The partnering programmes set out below are based on typical scenarios:

A single project; not forming part of an alliance or framework; six months in design and fifteen months in construction; handing over in month 21 (a programme for this scenario is shown in Fig. 21.1):

- ❑ the initial partnering workshop will be held in month 1 followed closely by the initial value and risk management workshops
- ❑ continuous improvement workshops will be scheduled quarterly for months 4, 7, 10, 13, 16 and 19
- ❑ after completion and handover in month 21 the post-project review will be held in month 24.

An alliance or framework, contracted for five years with multiple individual contracts:

- ❑ the initial framework partnering workshop will be held in month 1
- ❑ full team training will, if appropriate, be held at about the same time
- ❑ a continuous improvement workshop will be scheduled for month 7 and may also take place at other times during the framework

	Design	Construction	Post Project
Initial Partnering workshop	●		
Value Management workshop	●		
Risk Management workshops	●		
Continuous Improvement review	●	● ● ● ● ●	
Post Project review			●

Figure 21.1 Programme of project workshops.

- ❑ annual reviews (incorporating continuous improvement workshops) will take place in months 13, 25, 37 and 49
- ❑ a post-framework review should be held in month 63 to ensure all issues have been addressed and all objectives met.

Within the alliance or framework, each single project (four months in design; eight months in construction; handing over in month 12):

- ❑ an initial partnering workshop in month 1 followed closely by the initial value and risk management workshops. One of these may be held on the same day as the initial partnering workshop or on the following day if the team is conversant both with each other and with the principles and processes associated with value and risk
- ❑ a continuous improvement workshop should be scheduled for month 9
- ❑ a post-project review in month 15 after completion and handover in month 12.

We acknowledge that workshops are resource hungry, so every effort should be made to maximise the output by preparing in advance and making the experience as enjoyable as possible for the team. This will motivate the team members and assist them in producing quality output.

Whilst hiring a venue may appear extravagant when boardrooms are available the use of quality facilities, away from the offices of those organisations concerned, has many benefits. Over the years we have facilitated workshops in hotels and in the offices and site huts of many organisations. We can confirm that the quality of output from workshops is directly proportional to the suitability of the venue. A suitable venue drives a successful workshop where every team member chooses to participate with energy and a positive attitude because they are comfortable and feel valued. A quality, neutral venue ensures that:

- ❑ team members interact during breaks and don't disappear to their desks, for example to check emails
- ❑ interruptions are only for the one really important message
- ❑ there is evidence of a greater respect for people and value for the team

- ❑ there is a whole team focus on the workshop rather than organising the sandwiches and clearing away the teacups
- ❑ all facilities (flip charts, projectors, pens, paper, etc.) are provided
- ❑ there is an adequate provision of power outlets
- ❑ the room size and layout are appropriate as specified in advance, leaving the facilitators free to concentrate on the workshop preparation rather than clambering over old flip charts, chairs, etc. and rearranging boardroom tables against the wall.

Before the workshop, the facilitators should ensure that the venue is laid out appropriately. Our preference is for no tables, as tables are a barrier to open communication and we want to encourage openness. The absence of tables also enables the team members to move around easily into various groups as appropriate during the day. Cabaret style layout also facilitates group work but boardroom layout should be avoided if at all possible as it is liable to set up confrontation.

The facilitator should inform all delegates of the date and venue as soon as possible and send a formal delegate pack two weeks before the workshop, containing:

- ❑ agenda
- ❑ details of the venue
- ❑ list of delegates
- ❑ relevant project information (budget, start and handover dates, etc.)
- ❑ questionnaire as appropriate to elicit successes, opportunities and responses to performance (for example, against the partnering charter).

The questionnaire should include a specific return date for responses and details to enable return by email, fax or post.

One week before the workshop the facilitator should analyse all quantitative and qualitative responses to the questionnaire and prepare material for the workshop.

The workshop report and a one-page executive summary should be issued to all team members within three days of the workshop.

A structured programme of team workshops and social events, involving all team members and held in appropriate venues, will help to build the integrated teamworking ethos. Developing and communicating this programme early in the relationship will demonstrate the commitment of the core group to making partnering work and provide the opportunity to monitor the progress of the relationship, driving continuous improvement to maximise value for all parties in the partnering and integrated team.

22 Initial Partnering Workshop

The purpose of the initial partnering workshop is to build an integrated team across the separate partnering organisations, ensuring common understanding and breaking down any silo culture. The agenda for the initial partnering workshop is built around the three key features of successful partnering relationships identified in *Trusting the team* (Bennett & Jayes, 1995):

- mutual objectives
- problem resolution
- continuous improvement.

The initial partnering workshop should be held as soon as practical after the various organisations have been selected to form the integrated team. The exact timing will depend on the procurement route but the earlier it is held after the team has been appointed, the greater the benefits. At this early stage there is more opportunity to make effective use of the exchange of information and input of expertise from client, consultant, constructor, specialist, users and other interested parties.

In projects that are procured through a single-stage tendered route, the workshop should be held as soon as the constructor has been appointed. The date for the workshop should be set in advance of tendering and should be notified to all tenderers in the bid documents, together with the roles of those expected to attend from the constructor organisation.

Where the selection is a two-stage process, the workshop should be held immediately after the first stage selection and before designs and costs are worked up. Similar pre-appointment information should be given to all prospective partners.

In the event that the constructor and consultants have been appointed to a programme of work, for example, a strategic framework, the initial strategic partnering workshop should be held as soon as the constructor and consultants have been appointed. Initial project partnering workshops should be held as each project reaches the stage of commitment to invest, in order to communicate strategic aims, objectives and processes (including strategic key performance indicators) to project teams.

We are often asked who should attend the initial partnering workshop. There is no definitive answer but the team should consider the following in arriving at the attendee list:

- ❑ The optimum number of attendees is between 15 and 25 although we have facilitated successful workshops with as many as 50 and as few as 12 attendees.
- ❑ The decision makers from all organisations should be present to demonstrate commitment.
- ❑ Invitees should include the client, client representative, constructor, design and cost consultants and key specialists.
- ❑ Invitations should also be extended to other interested parties, especially those who will be impacted by the delivery of the project – for example, auditors and finance to input to sustainability and life cycle issues, users and maintenance representatives to input to operation and use, elected members to demonstrate support for partnering as the appropriate procurement route to best value.

In some cases, the team will include interested parties who are not used to the demands of an eight hour day spent on what may appear only to be technical issues. The facilitator should bear this in mind in structuring the day and ensure the involvement of all who have given their time. This can be achieved through small group working, regular breaks for social interaction and group exercises (for example, to demonstrate the value of teamworking).

Group exercises can be more or less serious according to the chemistry of the team. A fun element is a valuable component of individual learning and team building. We are all aware that we have more energy when we are having fun – this applies in the workplace as well as at leisure. Also, when we are having fun we tend to be more creative. Teams should be encouraged to have fun and be supportive of each other.

> 'As you enter this place of work please choose to make today a great day. Your colleagues, customers, team members, and you yourself will be thankful. Find ways to play. We can be serious about our work without being serious about ourselves. Stay focused in order to be present when your customers and team members most need you. And should you feel your energy lapsing, try this surefire remedy: find someone who needs a helping hand, a word of support or a good ear – and make their day' (Lundin, Paul & Christensen, 2000).

The agenda and process set out below forms the basis for our initial partnering workshops. The agenda will be sent to all delegates two weeks before the workshop, communicating the principal objectives of the day. These objectives are to deliver a partnering charter of mutual objectives, agree an issue resolution process, commence the drive for continuous improvement and build the integrated partnering team. It is possible that the core group, in conjunction with the facilitator, will add to these objectives and amend the agenda to suit the specific relationship or project needs.

09.00 Introduction

The facilitator should introduce her/himself, state the objectives of the day and ask each team member to introduce themselves, explaining their role and interest in the project. An icebreaker, at this stage, will also help break down barriers. Senior managers from the partnering organisations should present a short statement of their commitment to partnering and developing the integrated team.

09.30 Mutual objectives

The purpose of this stage is to align the separate objectives of the partnering organisations in a signed partnering charter of mutual objectives. The facilitator will arrange the team in a number of organisation-specific groups (for example, client, constructor, consultants, interested parties) to identify a limited number of their corporate and personal objectives and share these with the full team. The team will then break into cross-organisational groups, tasked with writing a limited number of statements of joint objectives that reflect all the earlier corporate and personal objectives. Finally, the full team will work these joint objectives into a partnering charter of mutual objectives which should be signed by all present as their commitment to the project. It is likely that the development of the partnering charter will take up to two hours and will probably straddle the morning break at 11.00. This time is well spent as the team will now understand and be able to support each others' objectives.

12.00 Issue resolution

The purpose of this stage is to develop an issue resolution process by agreeing appropriate levels of decision making and defining equivalent roles in the different organisations. The facilitator will draw up an issue resolution ladder and elicit from the team the names and roles of those empowered to resolve issues at the various levels. In this stage the team may also identify the partnering champions or core group.

13.00 Lunch

This is a valuable time for social interaction and the facilitator should take note of the *buzz-level*, whether the team retracts into organisation-specific groupings, whether individuals stay aloof or even leave the venue at this time. The facilitator should prepare the charter for review and signing by the team immediately following the lunch break.

13.45 Team exercise

Following signing of the charter, the facilitator should introduce a team-based exercise to maintain energy levels, introduce an element of fun and help build the integrated team ethos. We normally allow between thirty minutes and one hour for this.

14.30 Continuous improvement

In commencing this stage, it may be necessary to take time to ensure a common understanding of benchmarking and key performance indicators (KPIs). If KPIs have been previously agreed, the facilitator should ensure that these are shared. If none have been previously agreed, the team should identify and agree a limited number of KPIs appropriate to the project. The team may then identify and work on specific continuous improvement opportunities using such tools as lean thinking and value or risk management. A break may be taken during this session.

16.45 Final presentations

The team will review the partnering charter of mutual objectives, the issue resolution process, KPIs and opportunities identified for continuous improvement. Actions should be reviewed and the team should agree the next steps, including the dates for future workshops.

The following outline is provided to assist facilitators in writing an executive summary of an initial partnering workshop (the words in brackets are guidance for the facilitator). The document should not exceed one A4 page and should cover all the salient points from the day including any actions but is not a replacement for the full report which may cover ten to twelve pages depending on font size, layout and content of the day.

The original objectives of the day were to deliver a partnering charter of mutual objectives, a structure for issue resolution and targets for continuous improvement. Further objectives were (insert any further objectives).

In developing mutual objectives, the team was arranged in organisational groups to determine their own corporate and personal objectives from the partnering arrangement. These were developed, through work in cross-organisational groups and full team session, into the following partnering charter which all signed.

(Insert the words of the partnering charter.)

In addressing issue resolution, the team developed an issue resolution ladder, identifying the project team members responsible for ensuring robust and rapid resolution of problems.

(Insert the completed issue resolution ladder.)

The core group consists of:

(Insert names of core group)

Project key performance indicators were agreed as:

(Insert KPIs)

In the continuous improvement stage of the workshop the team used the value/lean/risk management process to identify:

(Insert the results of the continuous improvement stage including actions)

Other actions identified were:

(Insert other actions agreed)

The objectives of the initial partnering workshop are not only to deliver a partnering charter of mutual objectives, agree an issue resolution process and commence the drive for continuous improvement, but also to build the integrated partnering team. The success of this workshop depends on the commitment of the team members and also on sufficient time being allocated to each of the objectives. We have found that it is essential to devote at least a full day to the workshop if all objectives are to be met.

23 Continuous Improvement Review

We have already mentioned that continuous improvement is the feature that differentiates a partnering relationship from integrated teamworking. In any integrated teamworking or partnering relationship there will be a need for the team to align objectives, resolve issues and meet on a regular basis to review progress, celebrate successes and address opportunities. However, a team that uses these reviews to drive further continuous improvement and manage knowledge for the benefit of all organisations in an ongoing relationship can be considered to be a full partnering relationship.

Continuous improvement reviews should be held at regular intervals throughout the life of the project, framework or strategic relationship to continue to drive enhanced value for all. If such reviews are not held there is the potential for team members to become complacent and settle into a cosy relationship where value is not added. The reviews should be held in an atmosphere that is challenging and energising and where the team members feel that they can make a difference.

Once the team has held an initial partnering workshop, the core group should consider the following as opportune milestones to bring the team together in a continuous improvement review:

- ❑ commit to construct
- ❑ start on site
- ❑ 30% through construction
- ❑ quarterly throughout the project or framework
- ❑ a major incident or success
- ❑ three or four weeks before specific key dates such as reports to executive or budget committees.

The dates and venues for the reviews should be agreed and planned in advance where possible so that team members can commit to the reviews.

There are many issues to resolve and opportunities to be developed in the early days of a partnering relationship, especially if team members are new to the concept of integrated teamworking. Our experience is that teams which commit to a structured programme of quarterly continuous improvement reviews succeed where others may not. Once the team is established and successes are rolling through the team, the frequency of these reviews may be reduced but this should be a decision of the core group of partnering champions, not a majority decision of the team members who may have tactical rather than strategic issues in mind.

Team performance and relationships should be constantly monitored. In order to measure team members' perceptions, we include in the pre-review delegate pack a questionnaire asking team members to score their perceptions of the team's performance against each of the points of the partnering charter. The scores are received by us before the review, averaged and analysed. Having facilitated many continuous improvement reviews, we have found that there is a consistent pattern to team members' perceptions of team performance against the mutual objectives of the charter in the early period of a partnering relationship.

The general trend in responses to the pre-review questionnaire is for high scores at the initial partnering workshop where enthusiasm is high. These scores are typically followed by a dip at the first continuous improvement review, once the team is involved in the day to day pressures of delivering the project. There is a danger that the wide range of opportunities identified at the initial partnering workshop develops into a long list of actions. In addition, directors may be looking for early results. Yet the benefits from the relationship take months or years to develop and quantify whilst the costs of reviews and facilitation are immediate. The core group should anticipate this dip and address it as soon as it becomes apparent.

In order to reduce the impact of the potential dip, the core group should limit the initiatives following the initial partnering workshop to a small number (say, three or four) that can be delivered by the team before the first continuous improvement review. The core group should also maintain their involvement in motivating the team and consider any incentives that may be necessary. Their proactive stance should reduce the impact of the potential dip in scores at the first review (see Fig. 23.1).

In the event that there is a dip in motivation and enthusiasm in the early months, the core group should take note and act immediately to turn this around. This is not the time to cancel the next continuous improvement review. Through the questionnaire, the team has been reminded of the charter that they developed in the initial partnering workshop. Taking a poll against the points of the charter has identified those areas which the team needs to address and has raised the team's awareness that something needs to be done. In the review, these issues can be aired openly in a no-blame environment and addressed rather than allowed to fester.

The benefit of maintaining a structured programme of regular continuous improvement reviews in the relationship is that the team can address the issues promptly and, just as importantly, identify and publicise their successes. These successes do not have to be major events, but initiatives that team members have worked on. The sharing of successes will provide the team members with evidence of progress and the team satisfaction that comes with recognition of jointly delivering added value.

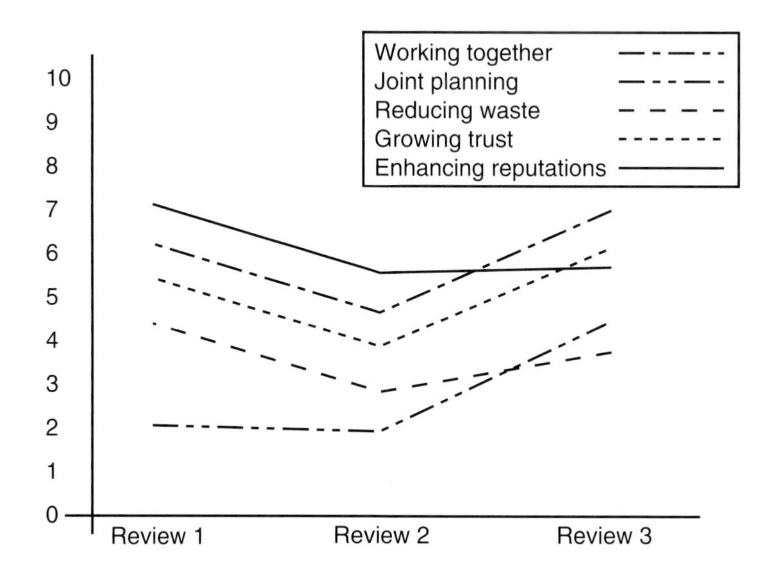

Figure 23.1 Quarterly charter KPIs.

The following brief report on success was drawn up by a cross-organisational group in a continuous improvement review for Sefton MBC Housing and their repairs contractors Integral. The team had identified the appointments process as a key area of success over the previous three months and we challenged them to produce one hundred words to encapsulate their success in this area.

'A pilot project for appointments for pre-inspections has been trialled in one area office and is now rolling out to other areas. This meets the objectives of the partnering charter by increasing tenant satisfaction and achieving best value through continuous improvement. Benefits have been secured for both tenants and the housing department through the elimination of abortive visits, the more efficient use of resources and from the convenience of tenants being able to select their own appointments. The contractors have also benefited from the reduction in variations and appropriately coded emergencies. It is intended to extend the scope of the service to contractor appointments offering even greater benefits to tenants, to contractors and to housing department staff.'

Note that this article not only summarises what has been achieved but also looks to the future. There is a drive to continuously improve and seek perfection.

Two weeks before the continuous improvement review, the facilitator should forward a delegate pack to all team members. This should include the objectives for the day, the agenda and a questionnaire to elicit each team member's perception of the team's performance against the partnering charter and a limited number of successes and opportunities for continuous improvement. A typical pre-workshop communication to all delegates is set out below.

Pre-workshop delegate pack for continuous improvement review

The objectives of the continuous improvement review – to be held on (date) at (venue) – are:

- to refresh and renew partnering for the team
- review progress on KPIs
- review progress on previous actions
- identify previous successes
- identify opportunities for improvement
- agree specific actions to drive continuous improvement.

At the initial partnering workshop on (date) the team developed a partnering charter.

(Insert the words of the partnering charter.)

Please show how you rate the team's performance over the course of the project so far by scoring each of the key objectives of the partnering charter. A score of 0 is poor and 10 is excellent.

(List each point of the charter separately for the team member to score.)

Identify two key successes in the project to date:

Identify two key opportunities for improvement over the next three months:

The agenda for the continuous improvement review will be as follows:

09.00 Introductions and objectives

09.15 Team based exercise to refresh the partnering and integrated teamworking ethos

09.45 Communication: in full session, including a review of KPIs and a review of actions from previous workshops and reviews. The team will also share the successes and opportunities identified in the feedback from the questionnaire.

11.00 Morning break

11.15 Observation: in full session, grouping the successes and the opportunities into key areas. The key areas will be listed, discussed and prioritised.

12.30 Lunch break

13.15 Learning: cross-organisational subgroups agreeing what to do to improve value for all, quantifying benefits and identifying who is best placed to implement the actions.

14.45 Afternoon break

15.00 Application: feedback from subgroups, proposing and agreeing specific actions to improve value

16.30 Review the day

A continuous improvement review is a good opportunity to hold a social event to cement relationships and build the team. This event may take the form of tenpin bowling, go-karting, a river trip or an evening meal after the review, all of which have been successful team building events organised by teams with whom we have worked.

The report on the continuous improvement review should be sent to all team members within three working days of the review. Some of the team may have been unable to attend the review but they should be included in the circulation of the report and of the executive summary which should summarise the review on one side of A4 along the lines of the example below (the words in brackets are guidance for the facilitator). We suggest one side of A4 in acknowledgement of fact that the executive summary is for a strategic overview of the review and resulting actions – the full report may run to ten or twelve pages of detail.

Executive summary of the continuous improvement review
The objectives of the review, held on (date) at (location), were to:

- ❑ refresh and renew partnering for the team
- ❑ review progress on KPIs
- ❑ review progress on previous actions
- ❑ identify previous successes
- ❑ identify issues and opportunities for improvement
- ❑ agree specific actions to drive continuous improvement.

The team reviewed progress on KPIs and discussed successes and opportunities arising from the data.

At previous reviews, (number) actions had been identified on the team to drive continuous improvement on this project. The team reviewed progress on these actions.

(Identify actions, the quantified benefits, any further work required, etc.)

The team shared the scores from the responses to the pre-review questionnaire, demonstrating the following trends in their perceptions of team performance against the charter.

(Insert graph of trends)

The team reviewed the success and opportunities for improvement identified by the team members in the pre-workshop questionnaire.

The key areas of success were:

(List key areas of success)

The key opportunities for improvement were:

(List key opportunities for improvement)

Arising from cross-organisational group work, the team identified the following new actions to drive continuous improvement in the remainder of the project:

(List actions to build on successes and actions to address opportunities, clearly and concisely identifying the actions, who is to lead each one, when each will be completed and the quantified anticipated benefits)

End of executive summary

The drive for continuous improvement is the differentiator between a partnering relationship and integrated teamworking. Regular continuous improvement reviews involving partners and all interested parties will enable team members to refocus on their mutual objectives away from the day to day pressures of the project, building on successes and addressing opportunities in a relaxed, solution focused environment. Continuous improvement reviews will add value for all individuals and organisations in current and future projects.

24 Post-project Review

The objective of the post-project review is to celebrate the successful delivery of the project and to share and spread learning within the project partnering team, driving continuous improvement for the benefit of all. Strategic partnering teams and frameworks will benefit from a structured process for continuous improvement beyond the current project, taking learning from the post-project and continuous improvement reviews into a central bank of project knowledge, whether this is a paper or electronic based system.

The post-project review is a review of the integrated team performance. Even though the project may be complete and handed over, each delegate remains part of the integrated team and should continue to view the project as a team initiative. The review is not held to assess the performance of any one organisation or individual (whether client, consultant or constructor) but should be an opportunity to reinforce the no-blame, solution focused culture within an environment conducive to individual and cross-organisational learning.

A post-project review should involve all disciplines and organisations who have worked on the project and include those who are taking over the occupation, operation and maintenance. A social event (for example, an evening meal) may be added to the review agenda to celebrate the successful completion of the partnered team project. It is also an opportunity to invite those whose efforts at the start of the project have contributed to the successful conclusion (e.g. estimators, groundworkers, etc.).

The format that we have adopted for post-project reviews is the same cross-organisational learning approach (COLA) that we use for continuous improvement reviews. It is inevitable that there is some duplication between this and the previous chapter but the agenda and examples within this chapter have been specifically adapted for post-project reviews.

The date for the post-project review should be agreed at the initial partnering workshop or very soon afterwards to ensure that diaries are prioritised. It should be held at a sufficient period after completion and occupation to allow time to complete the project and to ensure that all data is available to compute the key performance indicators (KPIs). However, this date should also be sufficiently close to handover that memories of the project performance are not dimmed. Somewhere between two and three months after completion is appropriate if a single post-project review is to be held.

Some organisations hold more than one post-project review in order to address different topics – for example the Open University New Library team held three reviews with the following objectives:

- two months after handover to establish the commissioning client's satisfaction and review how the project went for the partnering team
- six months after handover to establish the users' satisfaction with communication during the project, the decanting to the new building and their views on the facilities
- thirteen months after handover to review the working of the new building and the users' perceptions over a whole year.

The post-project review should incorporate:

- a review of the team's perception of their performance against the project partnering charter of mutual objectives
- a review of actions and measured benefits agreed in earlier reviews
- completion of the project KPIs
- identification and quantification of successes in the project that should be communicated as best practice to other partnering teams
- development of actions to enhance value to all partners on future projects.

In framework relationships, the development of a standard format for post-project reviews assists the teams in spreading best practices across the framework. This may lead, in time, to the same agenda being completed in a shorter period as teams become familiar with the process.

On completion of the post-project review (especially within ongoing partnering and framework relationships), the successes, opportunities and actions should be logged and stored in such a way as to enable easy retrieval in the early stages of future projects. Successes should also be quantified and the good news shared with interested parties whether or not they have been directly involved in the project.

Two weeks before the post-project review, the facilitator should forward a delegate pack to all team members. This pack should include the objectives for the day, the agenda and a questionnaire to elicit each team member's perception of the team's performance against the partnering charter, their identification of successes on the project and of opportunities for continuous improvement in future projects. A typical pre-workshop communication to all delegates is set out in the following box.

The objectives of the post-project review to be held on (date) at (venue) are:

- ❑ to celebrate the success of the integrated teamworking approach to the project
- ❑ finalise calculation and presentation of KPIs
- ❑ close out all previous actions
- ❑ identify successes of the project
- ❑ identify opportunities for improvement
- ❑ agree specific actions to drive continuous improvement in future projects.

At the initial partnering workshop on (date) the team developed a partnering charter:

(Insert the words of the partnering charter)

Please show how you rate the team's performance over the course of the project by scoring each of the key objectives of the partnering charter. A score of 0 is poor and 10 is excellent.

(List each point of the charter separately)

Identify two key successes in the project:

Identify two key opportunities for improvement in future projects with this team:

The agenda for the post-project review will be as follows:

09.00 Introductions and objectives

09.15 Communication: in full session, to include the calculation and presentation of KPIs and radar chart. The team will also share the successes and opportunities identified in the feedback from the questionnaire and close out actions from previous reviews.

11.00 Morning break

11.15 Observation: in full session, grouping the successes and the opportunities into key areas. The key areas will be listed, discussed and prioritised.

12.30 Lunch break

13.15 Learning: in full session, quantifying and agreeing the value of benefits.

14.15 Application: in full session, agreeing the steps that need to be taken to improve value for all in future projects.

14.45 Afternoon break

15.00 Team based exercise to celebrate and reinforce the integrated teamworking approach

16.30 Review the day, thanking the team members and sharing such awards and rewards as have been agreed by the core group

The report on the post-project review should be sent to all team members within three working days of the review. Some of the team may have been unable to attend the review but they should be included in the circulation of the report and of the executive summary which should summarise the review on one side of A4 along the lines of the example below (the words in brackets are guidance for the facilitator). We suggest one side of A4 in acknowledgement of the fact that the executive summary is for a strategic overview of the review – the full report may run to ten or twelve pages of detail.

Executive summary of the post-project review

The objectives of the review, held on (date) at (location), were to:

- ❑ celebrate the success of the integrated teamworking approach to the project

- ❑ finalise calculation and presentation of KPIs
- ❑ close out all previous actions
- ❑ identify and quantify successes of the project
- ❑ identify opportunities for improvement
- ❑ agree specific actions to drive continuous improvement in future projects.

The team shared the data on all project KPIs and constructed a radar chart to show performance against the industry norm.

(Insert radar chart)

At previous reviews, (number) actions had been identified on the team to drive continuous improvement on this project. The team reviewed progress and closed out these actions.

(Identify actions, the quantified benefits, etc.)

The team shared the scores from the responses to the pre-review questionnaire, demonstrating the following trends in their perceptions of team performance against the charter.

(Insert graph of trends)

The team reviewed the success and opportunities for improvement identified by the team members in the pre-workshop questionnaire.

The key areas of success were:

(List key areas of success)

The key opportunities for improvement were:

(List key opportunities for improvement)

The team quantified the value of the successes and agreed the steps that need to be taken to improve value for all in future projects:

(Schedule the quantified successes and the steps to improve future value)

Following completion of the formal review, the team took part in a team exercise (identify the exercise) and the core group thanked the team for their efforts, sharing such awards and rewards as follows:

(List any awards and rewards)

End of executive summary.

Many construction projects drift along after completion on site until the final account is agreed some months or years after completion. There is rarely the drive or opportunity for the team to place a full stop at the end of the project and capture learning for future benefit. A structured post-project review enables the core group to do this and to drive the completion of KPIs and final accounts to a specific date, agreed early in the project life. This gives an impetus to effective termination of the project with the full team, including users and other interested parties.

25 Value Management

Value management (VM) should not be confused with cost cutting. The latter can be carried out by an individual working in isolation and making adjustments to specification, quantities, etc. in order to bring a project or element within a predetermined cost limit. Working in isolation, the project member may not have full knowledge of the impact of this work on other areas of the project and may therefore remove critical requirements – 'throwing the baby out with the bathwater'. Value management, on the other hand, is an integrated full-team approach to identifying the needs of the project, proposing and developing alternative ways of delivering these needs at the appropriate price.

In the earlier chapters we discussed the need for clients, consultants and constructors to identify their value criteria. Value is defined by the person or organisation who pays, not by the organisation that delivers the product. We each determine which car, tv or house delivers the best value to us and our value criteria may change as our circumstances change. It is inappropriate for the car, tv or house salesperson to second-guess our needs. The European Standard on value management defines, 'value' as 'satisfaction of needs' divided by 'use of resources'. An alternative definition is 'required functionality' divided by 'life-cycle cost'. Using either formula, value can be increased by reducing costs (or resources) or by increasing the satisfaction of needs for a given cost.

The stages of the value management workshop are designed to enable teams to identify costs and resources, needs and wants (see Fig. 25.1). The stages are deliberately set apart in order to separate logic from creativity and provide the team with a structured process

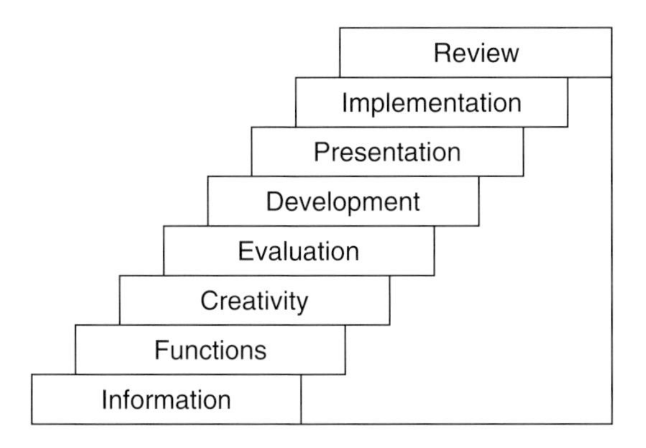

Figure 25.1 The value management process.

for reaching quality business decisions based on clearly defined value criteria.

1. information exchange
2. analysis of needs and required functions
3. proposal of alternative solutions/creativity
4. evaluation of alternatives
5. development of shortlisted proposals
6. presentation and consensus on action.

The final two stages of the structured value management approach are implementation of agreed actions and review.

Each workshop stage should be approached separately and the facilitator should ensure that, for example, creative proposals are not put forward during the exchange and analysis of information and that judgment on alternative solutions is not made during the creativity stage.

A VM scoping workshop may be held very early in the project life. Within a framework or strategic partnering arrangement this should include representation from the full integrated team and interested parties. However, in a single project that may yet have to be tendered to multiple constructors, this workshop may only involve the client and the cost or design consultants.

Within an integrated teamworking approach to a specific project, the first full value management workshop should be held as soon as possible after budgets and requirements have been outlined (for example, at the scoping workshop). The earlier in the project life that the team can get together, the greater the impact on adding value to the project. Later decisions tend to produce smaller proportional benefits, have a detrimental impact on the programme, cost more to implement and meet greater resistance. Thus, the earlier the VM workshops can be set up, the better – before the resistance and the cost to change exceed the benefits (see Fig. 25.2).

At the first full VM workshop, it is particularly important that the whole team is involved (including end-users and other interested parties). It is our experience that the presence of non-construction interested parties ensures that the discussion revolves around the user needs rather than technical construction detail. The facilitator should ensure that interested parties are actively involved and listened to by the technical experts.

As the project progresses, there may be a need for further VM workshops to explore specific functions in more detail (for example, the heating and cooling of the building). There is frequently an argument that a services or heating and ventilation workshop should not involve the end-user as this topic is too technical. However, in our experience, user-satisfaction of the heating system is critical to achieving user-satisfaction of the project. The benefit of an integrated team approach to VM is that the technical experts and users alike make

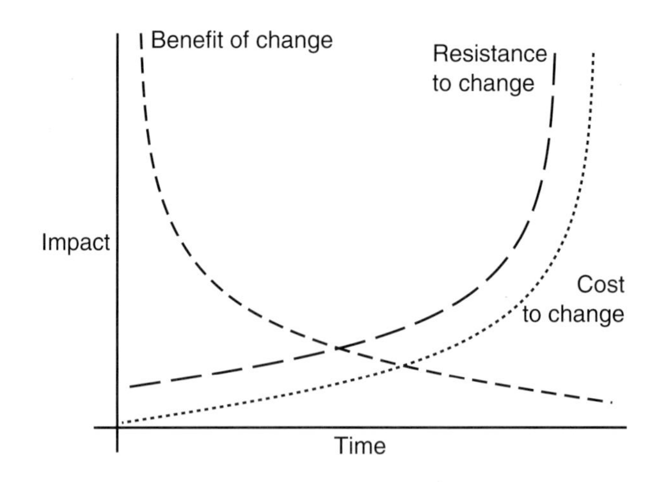

Figure 25.2 When to hold a VM workshop.

their cases for alternative proposals and the team buys into whichever option is best value, bearing in mind the stated needs of the project and the value criteria of the organisation that is paying. Arriving at these conclusions in a structured workshop in which the decisions are validated and recorded, ensures auditability of the decision making process and saves time and resource revisiting proposals later in the project.

In the outline agenda that follows we have assumed a one day value management workshop commencing at 09.00 and finishing at 17.00. Two weeks before the value management workshop, the facilitator (having discussed the workshop objectives with the core group) should forward a delegate pack to all team members. This should include the objectives for the day, the agenda and a request for team members to bring to the workshop all relevant information such as the brief, budget, cost plan, outline and detailed drawings and specifications. It will be of value for the facilitator to have received this information before the workshop in order to identify issues and prepare for the workshop.

09.00 Information Exchange

During this stage of the workshop the team will share information on project costs, programme, design, roles and responsibilities. The facilitator will seek to identify any misunderstanding across or between team members and ensure that the full team has a common understanding of the brief, budgets, quality and programme. As an example of potential misunderstandings, we have had occasions where we have been provided with four or five different cost figures by different members of the team. Some team members will include (and some exclude) VAT, some will include furniture and fittings or fees. Some team members will be very specific on costs and some will round the costs to the nearest £10 000. Time spent clarifying common understanding during this stage is critical to the success of the workshop.

The facilitator will note key information on a flip chart. This will help to reinforce understanding as all team members have clear sight of the information and can challenge assumptions.

By the end of the information exchange stage, it is important that all the members of the full team have a common understanding of all the critical elements of the project and that each team member has had the opportunity to identify any issues.

10.30 *Analysis of needs and required functions*

The second stage of the VM workshop is, in our opinion, the stage that separates value management from other cost reduction techniques and the process that adds value to value management.

The traditional approach to costing a project is by pricing elements of the work and this presupposes that the element is actually required. Value management makes no such assumptions. The facilitator should start from the point of questioning *why* the project is going forward. By repeatedly questioning *why* and *how* the facilitator will draw out from the project team the needs and wants of the project, identifying required and redundant functionality and specification.

In the early days of VM, it was identified that the functions of a product or a project could be more precisely defined by using fewer words. The classic function definition has been distilled to just two words – an active verb and a measurable noun. However, the use of the definition does not finish there. Whilst we can easily price 'build wall' (and we can build that wall more cheaply) we should be asking ourselves what the wall does rather than what it is or how it is constructed. In order to achieve this, the facilitator will ask the team 'Why are we building the wall?'

The two-word (verb–noun) response to the question 'Why are we building the wall?' might be 'enclose space' or it might be 'support roof'. The facilitator will continue to ask *why* against each of these other functions and, in doing so, will construct a diagram in which the higher order functions (those which answer the repeated questioning *why*?) will be clearly separated and identified.

The USA approach is to position the higher order functions to the right of the diagram and lower order functions to the left. However, in the shorter workshops that we facilitate, we prefer to place higher order functions at the top of the diagram and the lower order functions at the bottom. This process works well with Post-its and with flip charts. We have included a simple example in Fig. 25.3, taken from the initial value management workshop for Octavia Housing and Care's White City project at which various departments of the client were represented, together with the full design and construction team of Como Group, Cartwright Pickard, Campbell Reith, B+C Delloye Architects, Atelier ten, Grant Associates and Calford Seaden.

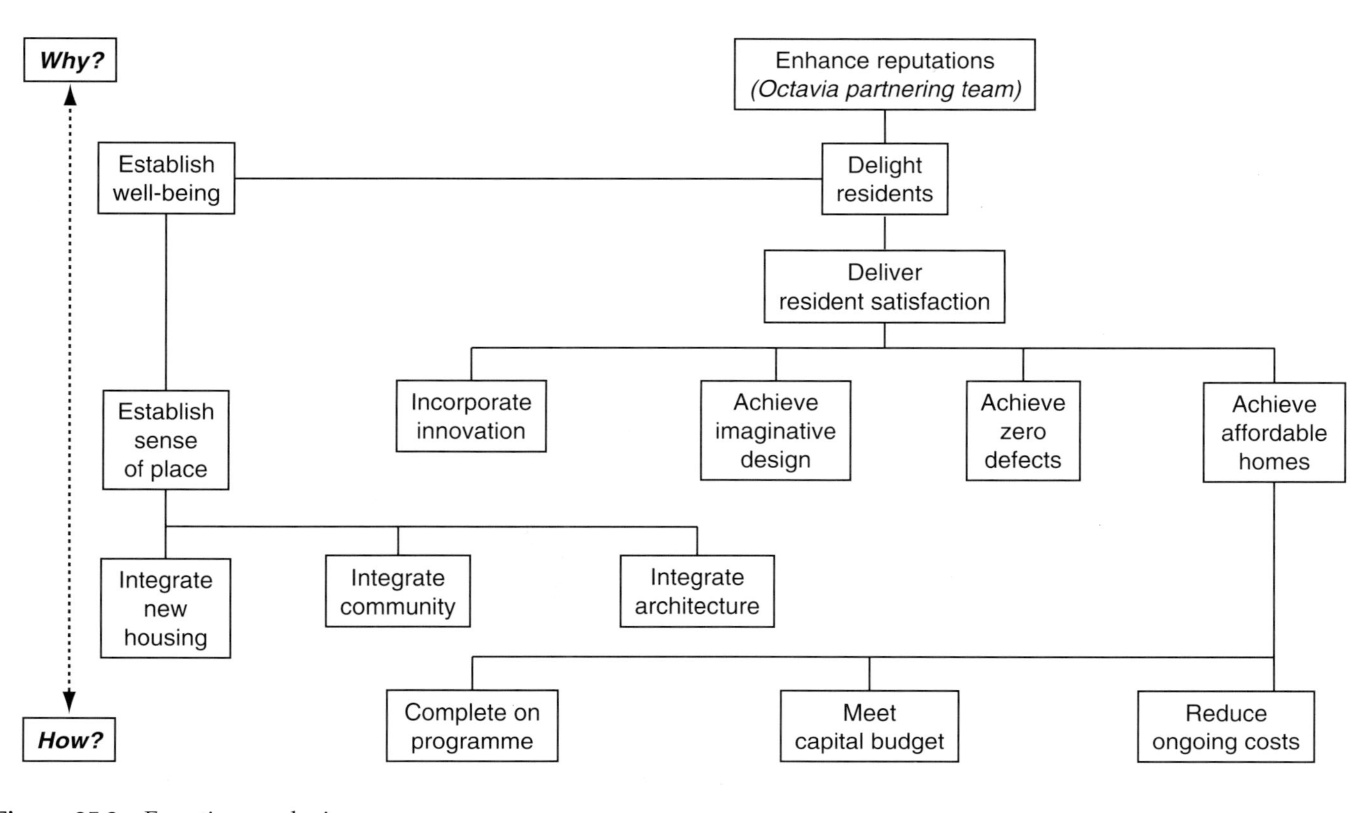

Figure 25.3 Function analysis.

11.00 Morning break

11.15 Creativity/proposal of alternative solutions

During this stage all members of the integrated team, regardless of their technical background, are encouraged to put forward suggestions for adding value to the project. The facilitator consecutively numbers and writes all proposals onto a series of flip charts and places these charts on the wall in full view of all the team.

The objective of this stage of the value management workshop is to give all team members the opportunity to put forward their suggestions for beneficial change without fear of recrimination or criticism. The facilitator should positively encourage lateral thinking in order to satisfy the stated needs of the project. All ideas are valid at this stage and comment on another team member's proposal is forbidden. If a team member disagrees with a proposal they may put forward a counter-proposal.

The facilitator will commence the stage by seeking alternative proposals to satisfy the higher order functions (for example, 'how can we deliver resident satisfaction?' rather than 'how can we reduce ongoing costs?'). This concentration on higher level functionality, with no challenges to any ideas put forward, encourages team members to be innovative and wide ranging in their thoughts and proposals.

It is not uncommon for more than 100 proposals to be listed in a half-hour session. Not all of these proposals will be seen to add value and some may seem inappropriate at first sight. However, they should all be noted on the flip chart and numbered consecutively for future reference. The team will be given the opportunity to discuss and evaluate all proposals in the following stage.

11.45 Evaluation

It is during this stage of the workshop that the team will jointly evaluate and prioritise those ideas that they believe add value and dismiss those that do not. The facilitator will lead the team through a discussion of each individual proposal raised in the creativity stage. The team will then identify whether each proposal is worth developing by scoring the proposal:

3 for a proposal with potential to add value
2 for a proposal which may have potential but about which the team may be unsure
1 for a proposal with little merit for this project and
0 for a wild idea

This stage will normally take twice as long as the creativity stage as there may be considerable discussion on some points. Some proposals may not be worth developing as individual items and may need to be linked to other proposals.

While the team is reviewing the proposals, a second facilitator may input the proposals onto a spreadsheet, together with the team's evaluation scores against the proposals (see Fig. 25.4). This will enable rapid sorting of the prioritised proposals, ready for the development stage.

12.45 *Lunch*

During the lunch break the facilitator will schedule the proposals which have scored 3 and 2, grouping these into topic areas. The facilitator will divide the team members into task groups and print out copies of task-specific proposals for the task groups (for example, all heating and ventilating proposals will be printed and passed to the services task group).

#	Creative proposal	Score	Linked to
1	Move kitchen from east end of building to west end, adjacent to incoming services	3	
2	Reduce area of kitchen from $70m^2$ to $50m^2$	3	
3	Omit kitchen and buy lunches from local burger bar	1	
4	Omit lunches	0	
	...etc...		

Figure 25.4 Creative listing.

13.30 Development

Each task group will develop the proposals relevant to their specialism, summarising their thoughts and their conclusions on standard sheets (see Fig. 25.5). These standard sheets will also prompt the groups to consider capital costs, life-cycle costs, programme, design, functionality, etc.

The team (and each task group) should prioritise development of the proposals that have the potential to add the greatest value to the project. Note that this added value can be by reduction in cost or by increase in satisfaction of needs or required functionality as defined in the second stage of the workshop.

The group will summarise the benefits and costs of each individual proposal, the advantages and disadvantages of the proposal and a recommendation to the full team whether or not the proposal should be accepted.

15.00 Presentation and consensus

In the final stage of the workshop, each group will present their recommendations to the full team who will discuss and reach con-

Creative Idea No___ Description	
Advantages of proposal	
Disadvantages of proposal	
Costs...Capital	Costs...Life cycle
Benefits...Capital/project	Benefits...Life cycle, Sustainability, etc.
Group recommendation:	
VM team conclusion:	

Figure 25.5 Standard VM proposal form.

sensus on each proposal, taking into account the needs and required functionality identified in the second stage. At this stage, it is necessary for the core group and any other key decision makers to be present to support the decisions of the team.

During this stage, the facilitator must be vigilant of the passage of time. All groups must be given the opportunity to feed back all their recommendations and there must be time left to review the actions and added value before closing the workshop at 17.00.

After completion of the workshop, the facilitator should ensure that the VM report is sent to all delegates and other team members as appropriate as soon as possible. We target three working days from the workshop for delivery to the delegates. This is made easier by email and by web-based collaboration software. The report should include a summary of the information exchange, a diagram of the needs analysis, the creative listing (including the scores and cross-referencing from the evaluation stage), and the agreed actions arising from the proposals with the quantified added value for each proposal.

Adhering to the structured value management agenda will demonstrate to all members of the team (including interested parties who may not have been present at the workshop), that all have had the opportunity to input proposals to add value and that all proposals have been recorded, considered and evaluated. The report will further demonstrate that those proposals with the potential for adding value have been further developed by the team and team consensus has been reached on implementation. It will be clear to any interested party that decisions have been based on objective value criteria, including consideration of whole life costs as appropriate to the project, and not on the individual opinion of a single team member working in isolation.

26 Risk Management

All organisations and individuals within organisations consider risks at some time during the life of a project. This consideration may then be followed by action, typically to avoid the risks or pass them to another party. Increasingly, clients and constructors are developing their own risk schedules for projects and taking action to ensure that the risks to their own organisations are reduced. Unfortunately, many such actions result in the risk being passed to another party without any consideration as to whether the other party is best placed, or even in a position, to take on that risk. Such actions, implemented in isolation, may create further risks for the wider project team and the project.

The separate development of risk registers has the potential for both parties allowing contingency for the same event, which may lead to over-budgeting and failure of the scheme to proceed. Alternatively, neither party may cover the risk, assuming that the other party has borne it and the project may proceed without due resource allowance or planning to cover the eventuality. In any case, this development of separate risk registers leads to duplication and ineffective use of resources and can cause confusion for all parties during the construction phase, when time is at a premium.

The development of an integrated teamworking approach to risk management reduces duplication of effort and increases all team members' understanding of the risks of the project failing to meet its time, cost and quality parameters. Through a structured process, team members can propose solutions which may not necessarily incur additional contingency sums or risk budgets. By sharing their thoughts, their experiences and their skills, the integrated team can significantly increase the likelihood of a project delivering to time, cost and quality criteria.

The structured risk management workshop has three clearly defined stages which are followed by the fourth stage – team management of the risks (see Fig. 26.1). In the first stage the team will identify the risks. This will be followed by an assessment of the impact and the likelihood of the event. In the final workshop stage, the team will plan together to mitigate the risks. After the workshop, the management of the risks is the ongoing responsibility of the team throughout the project, even though the management of the risk register may be in the hands of an individual project manager or client representative.

An early risk management workshop should be held very early in the project life with representation from current team members and interested parties. As with value management, the initial risk management workshop for a single project, that may yet have to be tendered to multiple constructors, may only involve the client and the cost or design consultants.

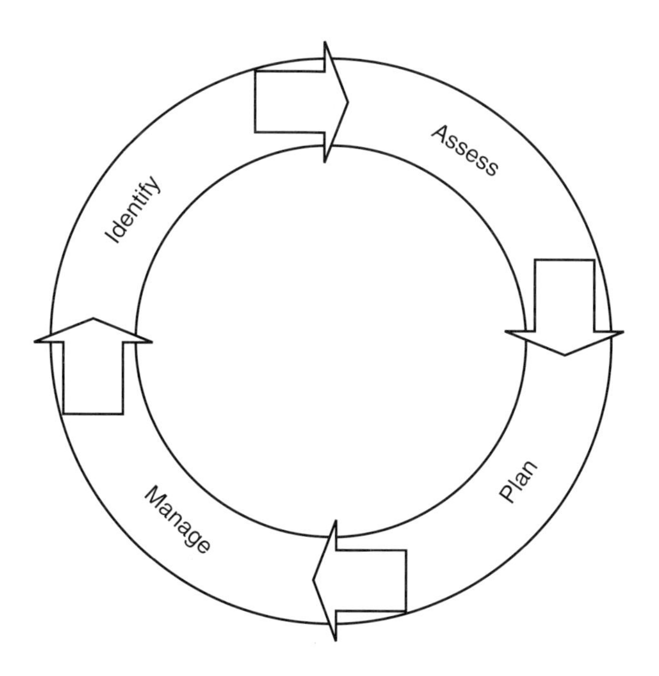

Figure 26.1 The risk management process.

Once the integrated team has been appointed and established, the first full team risk management workshop should be held as soon as possible after budgets and requirements have been outlined. Greater cross-organisational benefit will accrue from early involvement of the whole team.

In the outline agenda that follows, we have assumed a one-day risk workshop commencing at 09.00 and finishing at 17.00. Two weeks before the workshop, the facilitator should forward a delegate pack to all team members. This should include the objectives for the day, the agenda and a request for team members to bring to the workshop all relevant information including budget, programme and drawings.

09.00 Identify

In the first workshop stage, the team will identify risks of the project failing to meet its time, cost and quality objectives. The facilitator will elicit these risks from the team and write them on a flip chart. The intention of this stage is not to seek an unlimited number of risks but to clearly define the risk, stating the event and the consequences of each individual risk.

It is frequently suggested to us that, if all the risks are identified, the project may not go ahead. This is only partially true. If all risks are identified and the team places £x contingency against each risk, the project may well not go ahead. Therefore, it is incumbent upon the team to define and consider each risk very carefully and, for example, not just add £5000 for the possibility of steel arriving late.

The facilitator should seek clear definitions of each risk, including the event and the consequence of the event happening. As an example, it is frequently stated that *inclement weather* is a risk. In the UK, inclement weather is not a risk but a racing certainty. Even a spell of hot sunny weather is inclement for some operations. A definition of risk on a project should be as detailed as, for example, 'high winds in October whilst hoisting prefabricated modules causes delay to the project'. The 'high winds . . . whilst hoisting prefabricated modules' is the event and the 'delay to the project' is the consequence. In this way, the team can be more specific in their approach to mitigating the risk in the later stages. In a two-hour stage, before breaking for coffee, it is likely that the team will identify 40 or 50 such events.

11.00 Coffee break

During the coffee break, the facilitator may group similar risks under generic categories such as operational, interfaces, groundworks, programme or budget risks.

11.15 Assess

The second stage of the risk management workshop is to take each risk, assessing the likelihood of the event occurring on the project and the impact on the project if the event does occur. Some teams will seek to identify a percentage against the likelihood and the impact but we prefer a four-step scale that comprises negligible, low, medium and high for both likelihood and impact. It is important that the team agrees at the outset the range for each category. An example is shown in Fig. 26.2.

The team will assess and rate each risk for likelihood and impact (see Fig. 26.3). By allocating a score to each rating (high = 4, medium = 3, low = 2, negligible = 1), and multiplying likelihood by impact, the highest risks can be identified on a range from 1 (negligible/negligible) to 16 (high/high).

	Likelihood	**Impact on cost or time**
Negligible	Less than 1 in 100 projects	Less than 0.1%
Low	Between 1 in 100 and 1 in 20 projects	Between 0.1% and 1%
Medium	Between 1 in 20 and 1 in 5 projects	Between 1% and 5%
High	1 in 5 projects or more	Over 5%

Figure 26.2 Scales of likelihood and impact.

#	Risk event and consequence	Lk	Im	Rtg	Management action, by whom and when
1	High winds in October prevent craneage to fifth floor, leading to delay of two weeks.	M	N	3	
2	Delay in obtaining planning permission leads to project delay of two weeks.	H	L	8	
3	Delay in obtaining planning permission leads to project delay of two months.	H	M	12	

Lk = Likelihood; Im = Impact; Rtg = Rating

Figure 26.3 Risk identification and assessment.

13.15 Lunch

Many risk management workshops finish at this stage. However, it is our experience that there is value in keeping the team together to develop management actions in the next stage so that actions and risk transfers are not imposed by a single individual or organisation.

14.00 Plan

Having rated and ranked the risk items, the team should work from the top of the list and identify who will own each risk, what the team will do to prevent the event occurring and what the team will do if, despite their best efforts, the event does occur. The team should also identify whether any risk sum is required for the specific item and, if so, how much and who provides it.

It is probable that low-impact, low-likelihood risks will not be addressed but absorbed by the team. For example, 'Paint supplier goes into liquidation during the contract.'

Generally, high-impact, low-likelihood risks will be transferred to the organisation best placed to manage that risk. For example, the client may handle the risk, 'Budget holder and project sponsor leave the client organisation before the brief is completed.' In some cases, events in this high-impact, low-likelihood area may be transferred to insurers. This will lead to additional cost (the premium) which is irrecoverable – even if the event does not occur.

Low-impact, high-likelihood risks will require management action to address the causes or to find alternative ways of carrying out the work. For example, 'Steel prices rise by 5% within six months, leading to 0.5% increase in project costs.'

The team should be especially wary of any risks identified as high-impact, high-likelihood. In our experience, many such risks are in this category because they are not clearly defined (for example, 'adverse weather'). The facilitator should challenge the team over any such lack of clarity, even if this leads to additional items on the register. However, if the event is genuinely high-impact, high-likelihood, this is a risk on which the team should expend considerable effort in reducing either impact or likelihood. For example, 'discovery of asbestos in roof space leads to two month delay to project'.

During the workshop, or at core group meetings, a plan should be developed and agreed to manage the sharing of the risk budget in the event that specific risks do not occur or the time for that risk has passed (see Fig. 26.4). The core group should clearly identify whether

#	Risk event and consequence	Lk	Im	Rtg	Management action, by whom and when
1	High winds in October prevent craneage to fifth floor, leading to delay of two weeks.	M	N	3	Review offsite storage. Client Rep by 3 May.
2	Delay in obtaining planning permission leads to project delay of two weeks.	H	L	8	Monitor progress of approvals. Architect by 17 Feb.
3	Delay in obtaining planning permission leads to project delay of two months.	H	M	12	Identify probable requirements. Architect by 17 Jan.

Lk = Likelihood; Im = Impact; Rtg = Rating

Figure 26.4 Risk planning and management.

the team shares the saving in a pre-agreed proportion or whether the risk budget reverts to the individual or organisation who took the risk. Each major risk that carries a substantial budget may require separate agreement.

(Post workshop) Manage

All of the foregoing information should be entered into a risk register which should be reviewed on a regular basis by the team. Risk registers may be compiled in word-processor, spreadsheet or database format but, whichever format is adopted, they should be easy to use and update.

Once a risk register is compiled, the team needs to agree a process for reviewing all risks on a regular basis. The most likely way of dealing with this is for the project manager to own the risk register and to review the key risks with the team at site meetings and in continuous improvement reviews.

In the reviews of the risk register, it will be identified that some risks will have been eliminated and some will have passed. These risks will need to be removed from the register and appropriate adjustments made to risk budgets. Other risks may still be current and the team should review whether their mitigation strategy is still appropriate in the light of changing project conditions. Further risks may have been identified and the team should assess, plan and manage these as outlined above.

At the end of every project, the team should review all the original risks, the mitigation strategies and the impact of the strategies on the project. They should ensure that lessons learned are fed forward into future projects through the post-project review.

27 Lean Thinking

Lean thinking (Womack & Jones, 1996) came to the attention of the wider construction industry following the publication of *Rethinking construction* (Egan, 1998). Sir John Egan's Construction Task Force identified Lean as one of a number of processes that were being applied by leading clients of the industry to reduce waste. Lean construction is an extension of the lean thinking principle, a way to deliver more customer needs with less effort, in less time, with less resource and fewer defects.

Lean thinking has five stages:

- define value
- plot the value stream
- make the value flow
- pull
- seek perfection.

Whilst the words may be unfamiliar to a construction team, the principles behind them will be very clear.

We use the lean thinking structure within continuous improvement reviews, with the full integrated team, to identify and challenge practices and processes that have built up over the course of years. One of our icebreakers has been to ask teams to identify the biggest waste of time in their working day. Identifying this waste is the first step towards the team members applying lean thinking to eliminate the waste. The fact that they have identified the waste and are empowered to develop processes that not only reduce the waste of time to them but also add value to users or their organisations, is likely to add to the team members' ownership of the problem and to their interest in implementing the improved process.

In order to demonstrate how we apply lean thinking within continuous improvement reviews, we have shown below the team analysis of an apparently simple process that had been identified as a problem for tenants, surveyors and operatives – the call-out from a tenant to the call centre of a repairs service, to carry out a repair to the front door.

Stage 1 – Define value

After much discussion, the team decided that, from the point of view of the tenant, the best value service was one in which the tradesperson would arrive quickly and complete the necessary work in one visit, saving the tenant from having to be present for one or more follow-up visits. From the point of view of the client, the best value service would have to consider lowest appropriate price for a timely and right first time service. The constructor's value objectives were achieved by enabling them to provide the service at the lowest appropriate cost whilst making a reasonable margin, preferably in one visit.

Stage 2 – Plot the value stream

We split the team into three groups and asked each group to use Post-its to identify all the stages from the initial tenant's request through to completion of the job. We omitted the invoicing and accounting stages in this exercise although we suspect that there was much potential for reducing waste in this area also. Each group had a different understanding of the process and we combined all their ideas onto the one flip chart which we have reproduced in Fig. 27.1. This stage of the lean process is not looking to find solutions but to identify all steps in the process and highlight those that do not add value to the tenant.

Much laughter greeted one group's suggestion that the tenant would ring the housing director or the local paper rather than the call centre, but this was the experience of some of the team members. Perhaps the particular tenant had received low value service in the past and discovered that this was the way to make things happen.

The facilitator should be prepared to take time at this stage to challenge whether the rules and regulations of the organisation reflect what really happens in practice. Management and staff in many

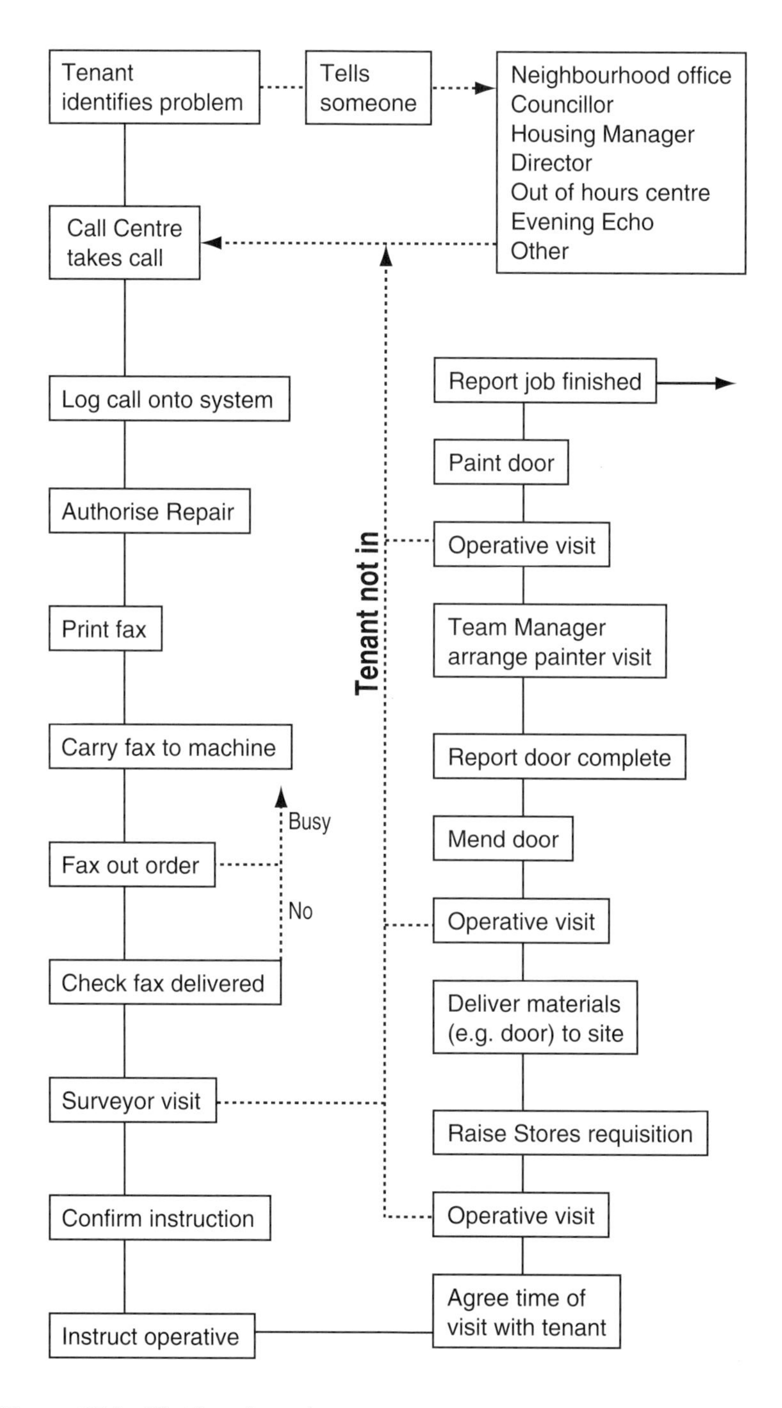

Figure 27.1 Plotting the value stream.

organisations spend much time devising ways of 'getting around' formal processes and it is this innovation and creativity that we are seeking to tap into.

In the process of plotting the value stream, the team also identified that there were four separate visits to site, all of which required the tenant to be in attendance if the job were not to be delayed. The chances of finding the tenant at home on four separate ad hoc occasions over a two-week period were deemed to be slight.

Stage 3 – Make the value flow

In this stage the team is looking to identify and remove items and processes that do not add value. One such process was the faxing of the order to the constructor.

The team identified that the constructor's fax machine was always engaged, leading to the client's operators having to wait and resend faxes. The team identified that this waiting time amounted to around ten minutes per fax. At ten minutes for ten faxes on each of five working days for fifty two weeks of the year this amounted to 26000 minutes wasted. A swift calculation converted this to 433 hours or 58 working days. At £100 per day total employment costs, this amounts to around £5800 of clerical time wasted every year. An email system was subsequently proposed by the team members and accepted by the management of the client and the constructor.

On the topic of repeat visits, the team agreed to look into the possibility of all visits (surveyor and operatives) being organised for the same time or to give the operatives authority to carry out work up to a certain financial level. This would avoid surveyors having to visit every job and would free them to carry out more proactive rather than reactive work, increasing their efficiency.

There are probably many more opportunities to reduce waste in this flowchart. The team certainly identified more and proposed more changes to the process to add value to tenants, constructor and client.

Stage 4 – Pull

In manufacturing, the concept of pull is to respond to the purchaser's immediate needs (they 'pull' the service or product). To do this, the team should make or do only what is immediately needed and not stockpile in store as storage is a cost that adds no value to the finished product.

It was interesting last weekend to apply the principle of 'pull' to a local burger bar. Whilst it may take ten minutes to cook and serve a single burger, it's fascinating to see how a really efficient burger bar team can serve you at the drive-in with full meals and drinks within a couple of minutes. How? There are burgers at all stages of production but they only move from one stage to the next when there is a need in the stage ahead. The operative at the counter pulls one from the rack and this is then replaced from the grill, where it is replaced from the kitchen, where it is replaced from the prep desk, where it is replaced from the freezer.

In replacing the fax process by email, the delivery of the instruction to the email box allows the constructor to pull the instruction when they have resource. The empowerment of the operatives to work to pre-agreed authority levels allows the operative to pull the surveyor to pre-approve or to check only when needed. This is an example of just-in-time rather than just-in-case.

An example of 'pull' in current construction projects is the increasing use of web-based collaboration software. Documents are revised and posted to the main server from which any person with a password can download the whole or part of the document as required, reducing waste of delay, printing and postage. Collaboration software also has the advantage that all team members are working from the same revision of the document at the same time, reducing the waste of working from out of date documentation and ensuring common understanding of the current situation.

Stage 5 – Seek perfection

Whilst the empowerment of operatives, the email system and the web-based collaboration software reduce waste and add value, there is still room for improvement. For example, instructions might be more accurate, response times quicker and document printing further reduced. So the team must constantly review its processes and procedures, challenging established thinking and attitudes and seeking perfection.

We find that using lean thinking processes in continuous improvement workshops is a rich vein in identifying and reducing waste, thereby adding value to all organisations. Regular use of the process helps to make lean thinking an integral part of the culture of the integrated team.

28 COLA – The Cross Organisational Learning Approach

This is an abbreviated version of a paper presented to the 6th International Conference of the Hong Kong Institute of Value Management (HKIVM, 2003) and subsequently reprinted in *Partnering for profit* (PSL, 2004).

Traditional, one-off, lowest price procurement of construction projects does not encourage the spread of knowledge or best practices. The adoption of project and strategic partnering by private and public sector clients of the UK construction industry has provided more opportunity for the management of knowledge within and across partnering teams. However, the gathering of information and transfer of knowledge within these initiatives has been haphazard.

The Cross-Organisational Learning Approach (COLA) was developed by a research team of academics, clients, consultants and constructors within the B-Hive DETR/EPSRC research project to structure the gathering and sharing of project knowledge within and across partnering teams. The use of COLA in a partnering and integrated teamworking situation adds value by improving the quality of feedback at all stages in a project, integrated teamworking or partnering relationship. The process assists teams to identify successes, address opportunities and increase organisational knowledge.

Since publication of the research on the B-Hive website (http://is.lse.ac.uk/b-hive/) in 1999, we have proactively promoted the use of COLA within partnering and integrated teamworking relationships and made further refinements to the practical application of the workshop methodology (see Fig. 28.1).

The COLA process provides a structure to extract explicit knowledge (the hard evidence in documents and reports, such as cost per m^2

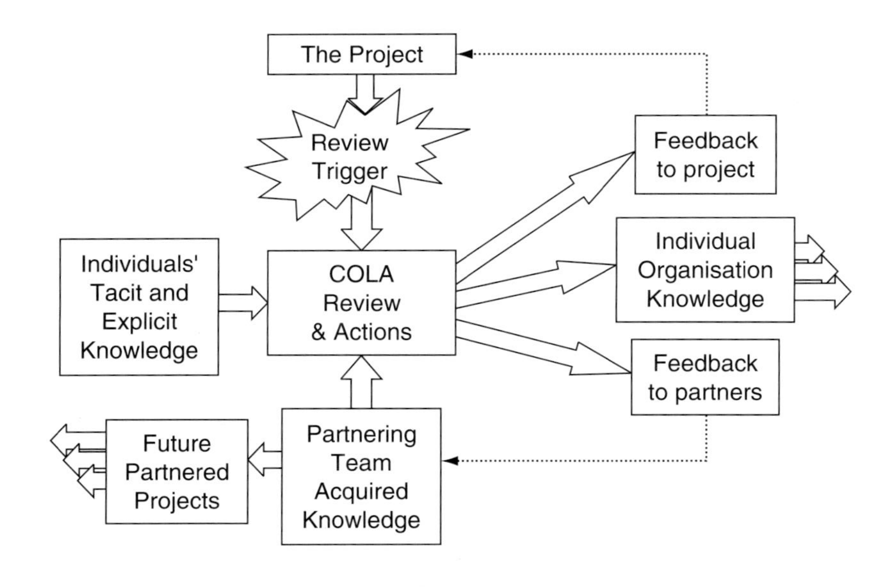

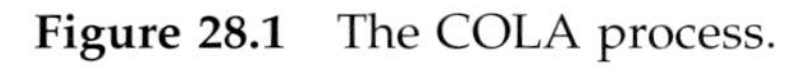
Figure 28.1 The COLA process.

or weeks to complete a project) and tacit knowledge (information held in the minds of the project team members, such as how well the team pulled together in adversity) from members of the team. The knowledge captured as a result of COLA reviews adds to the organisational knowledge of the partnering team and assists with driving continuous improvement on future projects.

Methods of storage and retrieval of the knowledge (whether by paper or electronic means) will vary according to the requirements and IT capability of the respective organisations. The summary of initiatives, successes and opportunities may be as simple as a word-processed document that can be searched for keywords, a spreadsheet or a bespoke database. The COLA project developed an electronic storage database but, whatever the system, it must possess three key features:

1. The knowledge must be accessible to all members of the organisations in the integrated team.
2. The knowledge must be maintained in an up to date state – the champions or the core group are the obvious custodians.
3. The knowledge must allow ad hoc entries from incidental learning such as comments arising at site meetings or informal discussions.

When a new project commences, the team should examine the database to ensure that past successes are repeated and opportunities addressed.

COLA comprises pre-workshop investigation and four workshop stages:

Communication – sharing successes and opportunities
Observation – listing, grouping and prioritising successes and opportunities
Learning – developing and quantifying the added value of proposals
Application – agreeing the owner and timeframe for the proposals and how success will be measured.

Two weeks before the workshop, the facilitator should forward a delegate pack to all team members seeking qualitative and quantitative responses to a series of questions. The form may ask for fact (explicit knowledge) and opinion (tacit knowledge) on various topics which might include:

- how well the team is coping with planning and phasing the project
- the team's performance in handling change
- the value being generated for individuals and their organisations
- cost, time and quality performance
- performance against project KPIs and the partnering charter
- inter-team relationships
- individuals' experiences
- successes and opportunities.

Forms may be standardised for consistency of comparison between projects or may be specific to a project, a phase or a specific problem. They should be tailored to be accessible to non-construction team members (such as social housing tenants, hotel managers and engineering operatives).

It is critical to the success of the workshop that there is robust representation from the whole supply team, the client, customers (end-users) and other interested parties. For example, a social housing project may include participation by tenants, housing management, board members, specialist constructors, best value auditors and legal departments as well as representatives of the client, consultant and constructor.

On receipt of the replies to the questionnaire, the facilitator will analyse the responses and consider key areas that might repay attention during the workshop.

Stage 1 – Communication

The objective of this stage is to elicit and share the explicit and tacit knowledge of the project team members, including the successes to date and opportunities for improvement in all key areas identified in their responses to the questionnaire. We have found that an effective process for communicating the knowledge of team members is for the facilitator to have written the questionnaire responses on Post-its before the workshop. This step preserves team members' anonymity and depersonalises the issues. The facilitator should use the words from the returned questionnaires, consistent with anonymity. Successes and opportunities should be displayed on separate flip charts. This generates discussion within the team, promoting understanding of the successes and opportunities. As an alternative, we have used a data projector to display the successes and opportunities in list form from a word processor package.

Stage 2 – Observation

Having communicated and shared experiences on the project, the team members group the Post-its into a limited number of key areas (or the facilitator does this grouping onscreen with input from team members). Following the grouping of responses, the full team discusses and prioritises the topics to be addressed in the next stage. It is important to ensure that the successes are not glossed over. Therefore, we encourage teams to write a paragraph or short report on each key success so that this can be cascaded to others, inside and outside the team as appropriate. In itself, this provides a valuable team building exercise.

Stage 3 – Learning

The objective of the third stage is to agree what to do to improve value for all and identify who is best placed to identify actions. This session is carried out in cross-organisational groups. Team members will choose to form or join groups to address specific key opportunities. The element of choice is important to obtain buy-in to decision making. Team members thus contribute to a topic that is within their area of expertise or on which they have strong views.

Each subgroup is tasked with identifying the appropriate steps to be taken to address the opportunity. They should consider criteria such as capital cost, resource availability, design and programme impact, sustainability and other agreed/appropriate criteria. The facilitator should make clear to the team that their discussions should result in enhanced value and that this value should be expressed in terms that are acceptable to the full team members. In many cases, the value can be expressed in cash terms. The facilitator should encourage teams to develop their thinking along these lines. Such thinking aids the development of a business case to support expenditure required to implement the action. For example:

- fewer meetings, resulting in a reduction in management cost of £x
- fewer defects, resulting in less revisits at a cost of £y each
- fewer complaints, reducing stress on management and staff resulting in lower sickness, absenteeism and staff churn – each of which can be costed.

Project: Date:
TOPIC:
What has to be done?
Who will lead it?
When will it deliver?
What is the quantified added value?

Figure 28.2 COLA feedback form.

Standardisation of the groups' responses is improved by use of a standard form (see Fig. 28.2).

Stage 4 – Application

The objectives of the application stage are for the groups to feedback their proposals and for the full team to agree specific actions to improve value. They should agree the team member to manage the implementation and review of the initiative.

The COLA workshop report should be circulated within three working days to all members of the partnering or integrated team, whether or not they attended the workshop. In this way, the learning is disseminated and the knowledge base grows for all.

29 Implementing Best Value

It is critical to the delivery of best value that the integrated team has a common understanding of the client's value criteria and the outputs against which the team will be judged. The integrated team can then align their objectives and ensure that all team members are pulling in the same direction, not cancelling out each other's efforts but utilising the collective intellect and resources of the team to implement best value (see Fig. 29.1).

The culture associated with delivering projects to lowest tendered price (the sum paid by the customer) will probably result in an attempt by the supplier to carry out the work at the lowest cost (the sums expended by the supplier). As a consequence, the client is likely to increase the level of checking to ensure that work has been carried out to required specification and verify that the supplier's lowest cost approach has not led to a reduction in standards. Any reduction in standards (whether of specification or quality of workmanship) will lead to a demand for rework and subsequent rechecking. All of these stages (finding a low tender, working to low costs, checking work, ordering and carrying out rework with additional checking after the

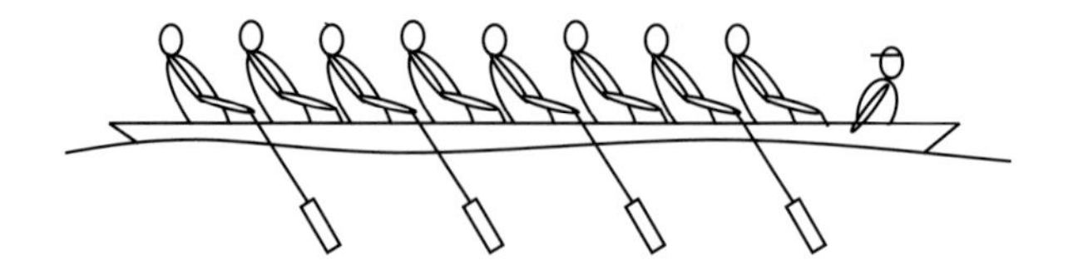

Figure 29.1 The integrated team.

rework) add resource cost to the work that may add no value to the client as the specified item is eventually delivered at the contracted price. Value will, in fact, be lost as the client will pay more than the contracted price (the cost of the checking resource).

Some clients will be able to identify their value criteria and share these with prospective partners in advance of the selection of the integrated team. The principles of best value apply for public authorities and the government defines this as,

> '...the optimum combination of whole life costs and benefits to meet the customers' requirement. This approach enables sustainability and quality to be taken into account...whole life costs allows factors such as fuel efficiency and replacement cycles to be taken into account, as well as social (e.g. benefits to local people, good workforce management, community safety, diversity and fairness). Successful procurement strategies are likely to be based on whole life cost considerations that include subsequent revenue implications and not simply the lowest tender price' (ODPM, 2003).

Public authorities are required to implement their services to clear standards of cost and quality (value) against which they publish reports on annual performance. In addition, they must review their services every five years to ensure that they are applying best value principles, not buying solely on lowest price but demonstrating that they have applied the *Four Cs* of best value to the service. The four Cs of best value are:

- ❑ Challenging why and how the service is provided (note that this requirement follows exactly the needs or function analysis phase of value management).
- ❑ Comparing performance with others (for example, using appropriate benchmarking and key performance indicators).
- ❑ Competing and embracing the principles of fair competition in deciding who should deliver the service (thus authorities will select suppliers on the basis of competition on robust value criteria which might, in addition to price, include evidence of a commitment to working cooperatively).
- ❑ Consulting local users and residents on their expectations about the service (involving interested parties in identifying their needs and the required functionality of the service).

In order to eliminate the redundant costs – those that add no value – the client must first be encouraged to identify their value criteria and share them with the integrated team in partnering documentation and at an early value management workshop. For example, does the client apply whole-life costing to their projects and, if so, how is this calculated and what discount rates are applied to future cashflows? Is sustainability a value criterion and, if so, what is the approach to calculating payback on sustainability proposals? Once the team understands the value criteria of the client, and the client understands the objectives of the other members of the integrated team, then the full team can take positive steps to add quantifiable value.

Value enhancement may not be achieved without a change in the culture of the organisations and the individuals within the integrated team. Habits and practices that the team members have developed, either consciously or subconsciously, over many years of working in a culture that rewards confrontation, will need to be identified and may need to be adapted to suit the new culture of cooperation. For example, traditional approaches have focused on reducing prices without always considering whether this reduces overall quality and thus value. A value-based culture will address needs and functionality at the same time as considering the costs and margins that make up price. Having achieved a change in procurement culture from lowest price to best value, the team needs to ensure that the other cultural aspects of all the partner organisations and individuals match each other.

Members of integrated teams may fail to recognise the adversarial language or culture of their current environment. In this event, a facilitator is able to hold up a metaphorical mirror to team behaviour and to challenge whether the behaviour is suited to an integrated team environment. For example, are team members using the language of blame ('why did you do that?'), concentrating on *my* project rather than *our* project or failing to act reasonably and without delay? Where there is a perceived need for cultural change, the team should be prompted to identify and develop behaviours that are appropriate to an integrated team. This should be addressed through raising awareness, joint training or, in extreme cases, removal of an individual or organisation from the team.

Partnering and integrated teamworking assists in the delivery of best value as the integrated team will be brought together early to understand each other's value criteria, to input their expertise at the most effective time and to work together, reducing duplication of effort for the benefit of the project.

The alternative option to the integrated team is the disintegrated 'team'. If a client had developed an integrated team would they break it up for any perceived benefits of disintegrated teamworking? The option of using a disintegrated collection (not a 'team') of specialists, consultants and constructors is likely to drive:

- rework
- checking
- lack of trust
- duplication of work
- multiple contracts and interfaces
- no drive for continuous improvement
- no benefit from learning curve and feedback
- all organisations focusing on their own goals and selfish objectives
- confrontation and escalations to claims rather than resolution of issues before the event
- poor or multiple-interface communication channels as shown in diagrammatic form in Fig. 29.2.

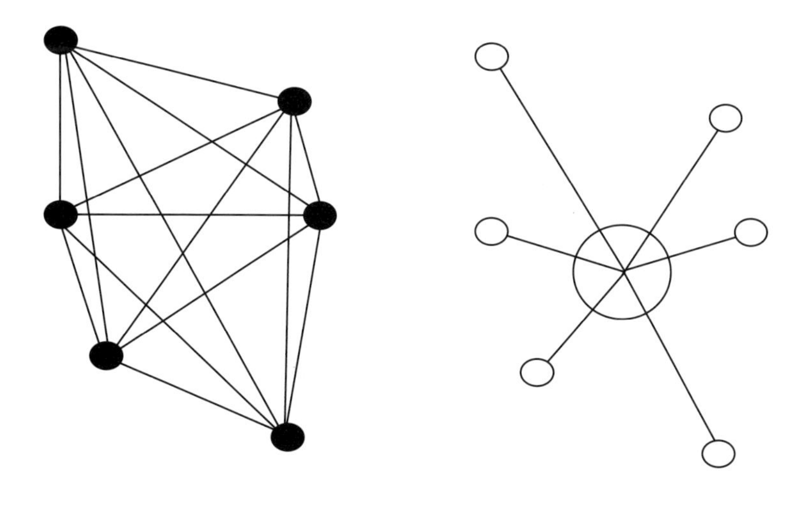

Figure 29.2 Disintegrated and integrated communication.

Sir John Egan's Task Force identified in *Rethinking construction* (Egan, 1998) that clients are generally dissatisfied with the output from a disintegrated construction industry. Without a focus on developing integrated teams, clients obtain lower value. They obtain a lower quality of service and product and pay for all the wastes listed above.

Integrated teams will deliver best value to all organisations when all team members are pulling in the same direction and not cancelling out each other's efforts. This will come about through a common understanding of value and a proactive effort to drive efficiency. In turn, this will make all team members' working lives more productive and rewarding, delivering best value not only to the organisations but also to the individual team members.

30 Sustainability

The World Commission on the Environment and Development defined sustainable development as '...development that meets the needs of the present without compromising the ability of future generations to meet their own needs' (Brundtland, 1987). A sustainable building project will add value by reducing waste during construction and operation, optimising resource efficiency, minimising adverse impacts on the local and global environment and providing a healthy environment for its occupants, improving their productivity.

The spin-off benefits of a proven sustainability approach will include a better corporate image for all organisations within the integrated team as well as the potential short and longer term cost advantages identified above. These are all quantifiable benefits that add value for clients, constructors, consultants, specialists and other interested parties and should be considered by the team whether or not there is direct influence from legislation or pressure groups.

There are three themes to sustainability – economic, social and environmental. Constructing Excellence encourages consideration of all three themes, guarding against undermining one section for the benefit of another and has developed a sustainability checklist for project teams in the stages of planning, design, construction and finished product (Constructing Excellence, 2003).

The Constructing Excellence environment key performance indicators at www.constructingexcellence.org.uk have been established to assist the industry benchmark and measure project performance, demonstrating benefits in:

- impact on the environment
- energy use
- mains water use
- waste
- commercial vehicle movements
- impact on biodiversity
- area of habitat created/retained
- whole life performance.

In addition, there are two secondary performance indicators (SPIs):

- all transport movement
- all transport distance travelled.

A problem that appears to face integrated project teams is that sustainability is a view of the future which they may not be able to address within the short duration of a project. We are regularly reminded in team workshops that capital budgets and programmes do not always allow a project team to consider sustainability, as the project has already been specified with a fixed capital budget, programme and specification.

In order to address this problem, sustainability issues – economic, social and environmental – should first be addressed by clients and their advisers in the initial stages of verifying the need for a project and assessing their options. During these stages, a 'no-build' option should be considered. As a result of a robust challenge, the client may consider that there is better value to be obtained from refurbishing or extending what is already in place.

When the decision has been made to go ahead with a project (whether new-build, refurbishment or extension), the client should explicitly identify their approach to sustainability in the brief – for example, whether or not they consider sustainability to be one of their value criteria. If sustainability is one of the client's value criteria, it should be one of the weighted criteria in the selection of potential partnering and integrated team members. The selection panel should score each supplier's submission on the basis of proven performance on appropriate sustainability issues.

Once the integrated team has been appointed, and assuming that sustainability is one of the team's mutual objectives, the team should jointly consider the three themes of sustainability as they develop the project, using the Constructing Excellence sustainability checklist mentioned above. Team workshops, using value and lean processes, will help to raise the team members' awareness of the benefits of considering sustainability within their project decisions.

As an icebreaker in a team workshop, the facilitator could consider using a quiz to raise the team's awareness of sustainability issues before they tackle specific project opportunities. We have included the Sustainable Development Indicators Quiz (DEFRA, 2004) within the chapter on team exercises.

During the project, the integrated team should consider appointing a sustainability champion to identify and record performance and feed back to the team at the end of the project, quantifying benefit as in the Llangefni example below. In frameworks or term contracts, the sustainability champion should be tasked with ensuring the spread of best practices by communicating successes around the wider team. It should be noted, however, that specific sustainability initiatives from an urban setting may not be appropriate for a rural community. Therefore, in the process of discussing how sustainability impacts on a specific project, the integrated team should work together to ensure a common understanding and to encourage proactive input of local knowledge from within and outside the team, especially the workforce. Note that, in the following example, the integrated team addressed all three themes of sustainability and used local specialist knowledge from within the team.

Early joint application of the principles of value management within a partnering arrangement enabled Cyngor Sir Ynys Môn (Isle of Anglesey County Council), JDM Accord and Hogan, to deliver stage one of the Llangefni relief road with enhanced functionality and sustainability.

Llangefni town centre relief road is a major element in the strategy to regenerate the town, enhancing traffic flow and providing better pedestrian facilities. The scheme is adjacent to, but outside, the Town Centre Conservation Area. Planning the programme together enabled the project to start within one week of confirmation of regeneration grants.

Environmental sustainability was enhanced. Local slate quarry waste was used as fill, reducing tips at quarries. Procuring this locally reduced the travel distance to less than 10 miles. The local demolition contractor broke up and crushed 9000m^2 of 200mm concrete slab, keeping this on site for fill, saving £70k and eliminating land fill tax. The new culvert had to be completed before migrating fish returned to this stretch of the river. Following a proposal from one of the workforce, the team set aside and reused the existing river bed gravel in the new culvert enabling migrating fish to identify the new route. Tidying the river banks, including removing obnoxious weed and concrete, improved the habitat and increased the potential for proliferation of different species of flora and fauna.

Demonstrating the team's commitment to social sustainability, extensive use was made of local organisations and labour from the Isle of Anglesey. The action of crushing concrete on site for fill eliminated the noise and dust of heavy lorries importing fill material and taking away crushings. This substantially reduced disruption to the people, traffic flow and businesses in Llangefni town centre. Initiatives to drive economic sustainability included the identification by the workforce of an existing, disused culvert which saved the disruption and £50k expense of creating a bypass channel. Standard kerbs and gullies in lieu of kerb drains saved £30k and reduced the likely cost of annual maintenance from £150 to £10 for each 20m length of kerb. As a result of the reduced town centre disruption, the council and contractors' management have a lower level of customer complaints and a lower incidence of health and safety, police and environmental health issues.

Construction teams are being forced to improve the environmental performance of their projects by legislation and through local, national and international pressure groups. Integrated teamworking can impact positively on sustainability issues in the planning, design, construction and finished stages of a project. Although a good deal of effort is involved, particularly in the early stages, the integrated team's concentration on sustainability can be of measurable benefit to all members of the integrated team and to the wider community, meeting the needs of the project without compromising the future.

31 Whole Life Costing

Forecasting and assessing the total costs of an asset over its whole life should be an integral part of any decision if the integrated team is to deliver the best value solution. A reliable approach to whole life costing relies on a clear definition of the life of the project (or the element) being considered and all the costs (including, but not exclusively, initial capital cost, renovation, repair, energy consumption, maintenance and disposal) incurred over that life. In addition, the team should understand how the client accounts for potential future cashflows, both in terms of income and expenditure.

In our experience, few integrated teams appear able or willing to address the issue of even a small extra initial capital cost for a saving in operating and maintenance expenditure, even if this revenue saving pays back the capital expenditure in a period as short as two years. If our experience is typical, and the concept of whole life costing is not widely practised, then integrated teams will be even less likely to address the further issue of sustainability. We believe that it is time for funders, accountants and auditors within client organisations to consider the required life of their buildings and the costs over the active life in the same way that they address the costs of their cars, taking into account purchase price, insurance, fuel consumption and residual value.

In training workshops we have set teams the following example of whole life costing:

A paint contract is to be let requiring 3000m^2 of painting with a capital budget of £30k.

Option A costs £10 per m^2 and requires repainting every ten years
Option B costs £5 per m^2 and requires repainting every five years
Option C costs £15 per m^2 and requires repainting every twenty years
Which option offers the best value and why?

Does it make any difference whether the project is a public sector library, a private sector high street restaurant or a public sector housing project?

Many teams immediately identify that the example is impossible to answer without further information on the reasons for the redecoration and the costs of disruption to the library, the regularity of scheme change within the restaurant or the average tenure of the rented accommodation. Yet team members may make such assumptions on major projects without adequate data or structured processes to support their whole life cost decisions.

One issue with whole life costing is that capital budgets are frequently (invariably) capped. No extra expenditure can be made, even if this results in a whole life cost benefit that pays back in six months, as the extra expenditure will be on the capped capital budget and the saving will be on somebody else's revenue budget. To counteract this, the Housing Forum made six recommendations to funders and the industry including alteration of funding systems, either for grants, loans or mortgages, to include consideration of initial capital and whole life costs (Housing Forum, 2002).

The topic of whole life costs should be addressed by partnering and integrated teams at the very earliest stage of the project life, once the decision has been made to commission a project. In the brief to the design and construction team, the client should set out the project approach to whole life costs to ensure common understanding. The information given to teams should set out clearly:

- the financial period over which whole life costs are to be considered (for example, is the team designing for a 30, 50 or 60-year life of the building and why?)

- the method of accounting for capital and revenue costs (for example, does the client take future income and/or expenditure into account in making a project decision?)
- the rate at which future cashflows are discounted.

> As an example of discounting future cashflows we have set partnering teams the task to consider how much they would take today instead of £100 in 12 months time. If they would take £90 their discount rate is 10%, if they would take £95 their discount rate is 5%. Generally, the higher the risk, the higher the discount rate that will be applied to future cashflows. So, for a very big risk, the team would take £50 now rather than £100 next year.

Armed with the facts concerning the anticipated life of the building and the impact of future cashflows, it is possible for an integrated team to make a reasoned business case comparing two alternative proposals. In addition to the information already provided by the client, the team will need clear data on all aspects of both alternatives including:

- the capital cost/initial cost to purchase
- annual costs such as energy consumption, regular preventative maintenance, warranties and service agreements
- annual repair or replacement costs, preferably based on past experience
- annual income, for example from rent or sales
- the costs and likely dates of removing and, if necessary, replacing the item
- the eventual costs of dismantling and removing the item at the end of its useful life, less any residual value
- grants, capital allowances and similar 'income'.

Acknowledging the need for sustainable solutions to construction problems, the team may decide to add a sustainability factor to their whole life cost deliberations. It appears that few sustainable initiatives produce an unarguable, cash-based, whole life costs benefit within a reasonable period (say, less than five years). However, if we accept that these initiatives have a sustainability impact and an

	Extra cost per home	Payback in years
High efficiency condensing gas boiler	£170	3
Improve insulation by 20%	£200	4
Combine heat and power plant	£1000	15
Solar water heating	£2000	18
Water saving devices	£120	25
Wind turbine electric generation	£3500	40
Photovoltaic panels	£5000	120
Grey water recycling	£1800	No payback

Figure 31.1 Payback periods for sustainable initiatives (Building, 2002).

added value to society, the addition of a pre-agreed sustainability factor to whole life costing exercises may be important. In such cases, the client may allow sustainable proposals to have a payback period two or three times longer than less sustainable alternatives (see Fig. 31.1).

Note that we have quite deliberately stated that the whole life costing exercise is within the remit of the integrated team and not solely of the cost consultant. In our view, integrated team decisions should be made by consensus and with the presence and input of all interested parties. Representatives from maintenance departments, auditors, elected members, board members and end-users should be present while major decisions are being made that will impact on their effective use of the facility. Their input is critical to achieving better value.

A paper on long term costs of owning and using buildings, published in 1998 by the Royal Academy of Engineering, was quoted in *Rethinking construction – an implementation guide for local authorities* (Green, 2000). The paper identified that for every £1 spent on the capital cost of construction, there could be far higher costs over 20 years in maintenance and in employing people in the building. A ratio of 1:5:200 was quoted for capital:maintenance:employment costs in a not untypical office building. Different buildings with differing uses could have other ratios, but all would show far greater costs over 20 years of occupation than for the initial capital cost of construction.

Thus, in order to ensure and demonstrate that best value will be (and has been) delivered for the substantial capital investments made in construction, the integrated team must forecast and assess the total costs of the asset over its whole life, taking into account not only the initial capital expenditure but also the impact on the owners, operators and users of the building.

32 Innovation

Innovation is the greatest opportunity for integrated teams, yet it can be the biggest turn-off. At one extreme, there are those who find innovation exciting and a real challenge. They will always be looking for more effective ways to achieve tasks, cutting red tape and reducing the amount of effort that goes into getting a result. At the other extreme are those who have developed systematic and workable procedures and who can see no benefit in changing what is a winning formula.

Some in the industry tend to think of innovation as the big step change, fantastic new product or new process. However, whilst innovation includes such step changes, it also includes the small changes that come from rethinking the way we carry out our tasks, through a different process or with different materials, in response to a problem or a customer demand.

The increased use of timber frame is frequently cited as an example of innovation, yet some very fine timber framed churches were being built by the Saxons more than 1000 years ago. Perhaps our industry shorthand is at fault here – timber frame isn't innovative but the modern use of timber frame is. Innovation may, therefore, include modern uses of old practices and this was brought to mind recently in a partnering review where the team had developed a community gang of tradespeople, empowered to relay paving slabs or carry out other minor necessary road and footpath maintenance in direct response to requests from members of the public. It was a practice that had been in use some years ago but, with the perceived need to approve all work at headquarters before expenditure, had been discontinued. The reintroduction of the process saved the delay and costs (around £30 per order) associated with ordering a small (say

£25) repair and gave added value to the customer as their concerns were being addressed speedily. The process was an innovative slant on an earlier practice.

Innovation is driven by the need to solve a problem and the desire to stay ahead of competition. Even winning Formula One teams continue to innovate in order to stay in pole position. The innovation comes from a culture inspired by the management, nurtured by their openness and their management techniques and supported by their close ties with their customers. Team members are empowered to solve problems and are highly motivated as a result.

The business case for innovation is that organisations that innovate for the benefit of their customers will overtake those who do not. But this is only true if the innovation adds value for the customer. So, if the integrated team want to introduce an innovative process or product, they have to make a very clear business case which convinces those who will pay for the innovation, implement it or use it.

Factors determining whether innovation is adopted include:

1. clear benefit – the customer must see a benefit in their terms, 'what's in it for me?'
2. familiarity – an innovation should be compatible with existing habits and practices making it easy to adopt
3. simplicity – simple is best as less effort is required
4. communication – the innovator should make it easy for the customer to understand the benefit in their own technical or business language
5. trial – involve the prospective customer in a low risk, low cost pilot.

Any process that helps the integrated team to question how they carry out their work and to propose beneficial changes will prompt innovative ideas. The value management process, for example, can be used as a tool for innovation. The structure of the value management agenda is such that it poses questions in the early stages (through sharing information and identifying required functions or needs) and then, in a separate stage, encourages the team to think laterally and identify creative solutions. Lean thinking also concentrates the team's collective mind on required outputs (the value to the customer). This process helps the team to analyse how the value is delivered (the value stream). Once the value stream has been plotted, the team will identify innovative solutions to omit non-value-adding stages, thereby reducing waste and delivering better value.

In our experience, creativity requires a relaxed mind and the ability to think laterally.

> 'Lateral thinking does not select but seeks to open up other pathways ... one generates as many alternative approaches as one can ... even after one has found a promising one ... generating different approaches for the sake of generating them' (de Bono, 1970).

In order to promote lateral thinking, the facilitator should create an environment in which the team feels relaxed and knows that all proposals will be valued. The opportunity to bounce bright ideas off each other will help the team in their development of innovative solutions. These solutions can come about as a complete surprise or as a gradual development of ideas and will be generated in greater numbers if the full integrated team, including end users and other interested parties, is involved.

The most innovative teams that we have worked with are those prepared to have fun together. They have their share of team members who will come up with the 'wild' ideas and the team piggy-back on these ideas. The really innovative teams tend to set us the challenge to provide a new team building exercise at every workshop and this fun element helps to break down barriers, aid learning and create an environment conducive to innovation.

If an innovative idea is identified at a workshop or through a suggestion scheme, a decision should be taken by management on whether it should be followed through. If the idea is to be pursued, management should identify a champion for the idea and set clear time and cost parameters within which a business case should be presented.

One might not think of a scaffold tie as an innovative proposal but it proved so in the context of modular construction. To Unite Integrated Solutions and their constructor partners, the ability to tie scaffold back to the installed modules using the original lifting plate has the benefit of speeding scaffolding and thus the overall programme. Modular construction is fast but the erection of freestanding scaffolding around the building was slowing the whole process. It was identified that each prefabricated bedroom module has a number of lifting plates fixed to the frame. After installation, these are redundant. In the course of a regional review workshop, one project site agent told the team that he

> had developed and trialled a threaded scaffold tie that would make further use of the plates if the lifting hole were also threaded. Previous proposals for using the plate to secure scaffolding had been complex, expensive (requiring gravity toggles) or difficult to remove. The team saw the benefit of the simple idea and identified that Unite Manufacturing should produce a plate with a threaded hole rather than the unthreaded lifting hole. The additional cost would be more than compensated by the saving in site time and an action was placed on the site manager to raise the proposal with manufacturing within five days.

Not all innovative ideas will be successful. We can all think of ideas that seemed good at the time but which failed to develop or to prove a financial success. A small proportion of research and development ideas make it through to successful product lines but without the drop-out ideas on which to piggy-back, the successful ideas probably wouldn't see the light of day. Teams should be aware of, and not be discouraged by, the relatively low implementation rate for innovative ideas.

If an innovative idea is not to be pursued, management should inform the team why this is so, as potential for future innovation will be stifled by the failure to develop or pilot a team's idea without giving a rationale. We have worked with teams who have produced innovative ideas which were accepted by the partnering champions on the day, yet were not implemented and no explanation was given to the team as to why this was so. The impact of this failure to communicate was to engender a spirit of cynicism within the team. This cynicism led to a reluctance to propose any such ideas in future.

Innovation is the key to continuous improvement. Partnering and integrated teams that innovate to add value for their customers will succeed where those that do not will fail. In order to convince their teams that innovation is not only necessary but delivers greater value to all individuals and organisations, the core group must provide a supportive environment (both in workshops and in day to day interaction) in which all team members feel comfortable that all ideas will be received and considered, even though only a small percentage may be implemented. In this environment, team members will generate innovative solutions to practical problems and this will keep the partnering and integrated team ahead of its competitors – driving added value for all.

33 Open Book Accounting

For the last ten years the construction industry has been encouraged to adopt cost and price strategies that are open book as far as is reasonable or practicable. It is argued that an open and honest approach to the costs and prices associated with a contract helps in building trust. Sir Michael Latham's draft report in December 1993 was entitled *Trust and Money* on the basis that the industry had, '. . . too little trust and not enough money' (Latham, 1993).

We have noted that some organisations have a reservation about the principle of open book accounting. In our opinion, this is due partly to a lack of clarity in the definition of the term. This is compounded by a lack of clear reasoning or mechanisms for applying open book principles. For example, we are working with one client organisation that has been told to, 'go open book' by their auditors, yet the auditors have offered no guidance on why or how this should be done, only that, 'open book will bring x% savings'. Open book may bring savings but there may be a commensurate risk with this and, before embarking on an open book approach, all organisations should consider carefully how a change to open book accounting may impact on them.

One definition of open book accounting is,

> '. . . a generic expression, which does not have an agreed definition, where each partner may agree to give the other a degree of access to his accounting data. The level of access, the manner in which it is to be delivered and the use to which it may be put must be agreed, on a case-by-case basis to reflect the circumstances of the partnering arrangement and the need for access to certain data to monitor performance or benefits arising' (MOD, 2002).

We are in full agreement with the principle of open book accounting if the costs and benefits have been carefully and diligently considered. It is our view that the core group should convene a meeting of senior management of the partnering organisations and appropriate interested parties (such as elected members and auditors) in full and open discussion at a very early stage in the partnering and integrated teamworking arrangement. Within this meeting, the representatives of all organisations should clearly identify and assess the benefits, costs and risks as they apply to their specific integrated teamworking arrangement before embarking on open book accounting.

It is important that the core group agrees and communicates to the members of the integrated team, the purpose of, and the value to be gained from an open book approach. They should set out their reasons for considering the option and quantify the benefits that open book accounting would bring to each party. Without quantification of benefit there is no business case for moving to an open book arrangement. The benefits of open book will differ according to the relationship and also according to the type of work being undertaken. For example, the benefits on a term contract for housing repairs may differ substantially from those on a single £5million new build project.

Generally, the more open book the relationship, the more cost risk there is on the client. This may be appropriate for a scheme where the extent of works is unknown and where budget control is secondary to safety or another key value criterion (see Fig. 33.1). For example, we were involved in an open book arrangement to repair a listed fermenting tower in a working brewery. The extent of the works was unknown but the budget was low priority compared with the need to keep the brewery working safely and effectively. Trust was built through regular meetings between the client and constructor both on site and through neither party letting the other party down at any stage in the contract. Trust grew as predicted cost, time and quality parameters were verified through very rigorous monitoring in the early stages.

Where budget control is a priority and the works (or elements of the work) can be clearly specified, measured and priced in advance, there may be less obvious benefit in an open book approach. The client may perceive a benefit in transferring the cost risk to the constructor as long as rates are fixed in advance and the client has

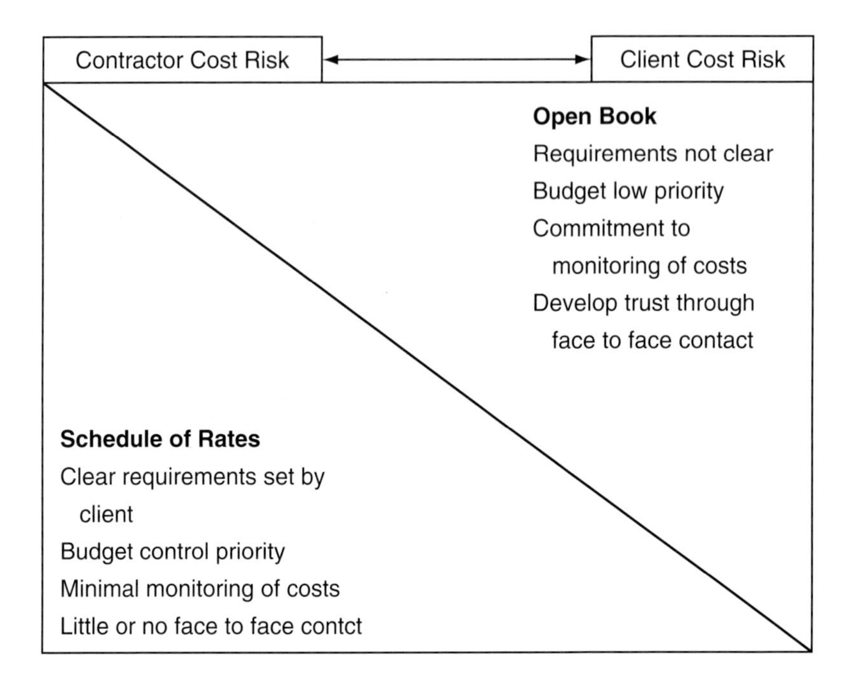

Figure 33.1 Open book options.

price certainty. For example, in a five year term contract for building repairs, the client may be prepared to pay extra in year one for the benefit of fixed prices over the term of the contract, passing the risk to the constructor.

The core group should note that effective application of open book accounting requires considerable trust between organisations and between individual members of the team, within and across organisations. Consequently, open book accounting may be best suited to organisations that have experience of partnering and a clear understanding of each others' value criteria and where levels of trust are higher.

After agreement on the purpose of open book and an assessment of benefits, the core group should address and define the level of access that each party will have to their partners' books and the resource costs associated with this. The client does not hold all the cards and may be expected to open their books in respect of, for example, total budgets and timing of future cashflows in return for access to the constructor's supplier invoices. The core group should clearly establish what level of information the client seeks from the constructor, what level of information the constructor seeks from the client and how much access each is prepared to allow the other.

Having established the purpose, benefits and the level of access, the core group should consider whether open book is going to lead to an increased level of checking, even if only in the early stages of the approach. Some interpretations of open book place reliance on a client representative checking a constructor's books after work is completed. This practice may not only be wasteful of resource but may also foster a spirit of distrust between the parties leading, for example, to discussions on whether the job really needed $5m^2$ of paving (as invoiced) or $4m^2$ (as measured net later).

As open book constitutes a major change for many organisations, the core group should consider whether there is an opportunity to set up the formal processes through trialling open book with a pilot project. Within the pilot, the team should review historical data to identify where issues such as cost and price overruns have previously occurred and assess where profit and risk allowances are accounted for in the schedule of rates and/or bills of quantities.

The integrated team should challenge specifications and processes through robust lean and value management processes. They should identify the open book costs (to the constructor and supplier) and separate these from the suppliers' risk pricing, overheads and profit. The risk pricing should then be the subject of a separate structured risk management exercise in which the team will decide, for each individual risk, whether it is most effectively placed with a supplier (including the constructor) or whether the client is best placed and prepared to take it on.

In developing their thoughts on open book, one project team discussed whether the procurement of a specific building element should be within the remit of the constructor or the client. A potential price reduction had been identified as being available to the client if specific building units were removed from the constructor's schedule of rates and bought direct. The joint team of client and constructor identified that the price saving came at potential risk to the client. The responsibility for supplying the units would pass to the client and any delay in supply would, on the basis of an open book approach, result in payment to the constructor for standing time. It was also identified that the client would have to increase their procurement resource in order to negotiate and purchase the units.

Working through the example enabled the team to jointly identify the risks and where these should lie with appropriate costs. On balance, the team decided that the current arrangements were an appropriate balance of cost and risk. The price difference was a fair price to pay for the constructor taking the risk of non-delivery on time. We consider that this exercise had been a good example of an open book approach as the team had identified and assessed costs and risk pricing separately from each other.

Finally, once the integrated team has agreed to follow an open book accounting approach, the implementation of open book should be treated as a project in its own right. It should have a project manager, an implementation team and reviews to assess whether it has added value for all partners.

There is little doubt in our minds that a carefully structured and objective open book approach to costs and prices, managed effectively by partners who understand each others' value criteria and who have built a degree of trust between organisations and individuals over a period of working together, will lead to better value for all concerned through greater awareness of the consequences of decisions and more confidence in costs and prices. This should lead to further development of trust and more opportunities to work together to add value.

34 Incentives and Rewards

For every £1 invested in a structured partnering approach, the benefits regularly exceed £3 in single project partnering and £10 in strategic relationships (Bennett & Jayes, 1995). These benefits provide the scope for incentives and rewards amongst the team. However, not all benefits delivered by partnering and integrated teamworking will result in cash that can be shared between the participants.

One of the key duties of the core group or the partnering champions is to consider the form of appropriate incentives and rewards at various times for the benefit of the project. An incentive is '. . . an inducement to motivate an organisation . . . or individual to place greater emphasis on achieving an objective or to act in a certain way' (Broome, 2002). Incentives should encourage collaboration and align the whole team's effort, focusing on more efficient ways to deliver the objectives of the team and the project. A reward, on the other hand, is recognition of performance exceeding expectations. Generally, therefore, incentives are placed before the event and rewards come later.

A major incentive for an organisation in any contract is to operate profitably. Traditional projects, in which each organisation is concentrating on the satisfaction of their own selfish objectives, frequently result in a situation where a win for one is at the expense of another. If this causes one or more team members not to be in profit, there is a further concentration by that organisation on winning the next battle and the relationship risks degenerating further into a series of win–lose confrontations. The integrated team, on the other hand, is looking for that elusive win–win situation (Covey, 1989) in which individual team members all gain as a result of the project being delivered cooperatively. If an organisation is assured of a profitable relationship, their concentration will be on delivery of the stated objectives of the project, rather than on seeking ways to make up potential losses for themselves. Thus, it is in the best interests of the integrated team to ensure that all team members make a fair return. This point can be demonstrated to the team by setting the red–blue exercise that we have set out in the chapter on icebreakers and team building exercises.

Traditional approaches to project costing favour a percentage fee or profit, based on the prime cost of the project. Ironically, if the team succeeds in reducing the overall price of the project to the client, the supplier (constructor, consultant or specialist) will lose fee or profit. This seems to us to be an inequitable way to do business but, more importantly, it seems to be a way that is almost guaranteed to concentrate the team members' minds on creating a high prime cost. If project teams were awarded contracts with fixed fees or profits, any reduction in the prime cost would lead not only to a reduction in the price but also to a larger percentage profit for the suppliers (see Fig. 34.1). This would be a major incentive to bring down prime costs.

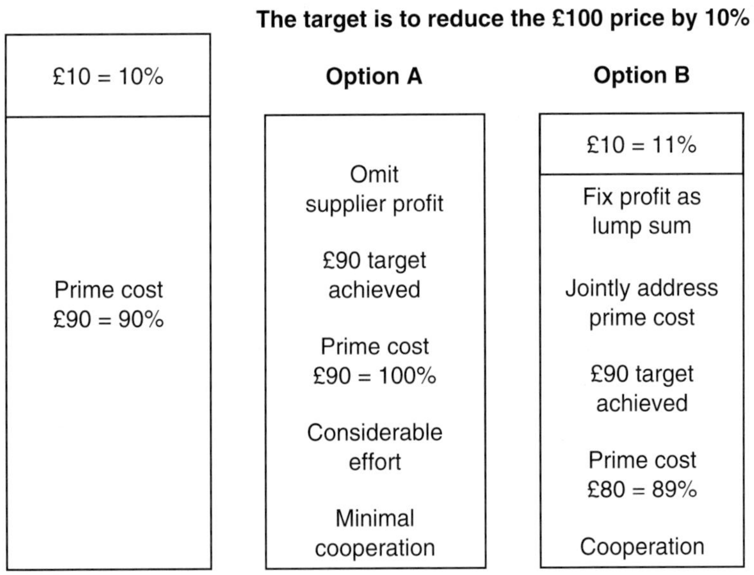

The target is to reduce the £100 price by 10%

£10 = 10%

Prime cost
£90 = 90%

Option A

Omit
supplier profit

£90 target
achieved

Prime cost
£90 = 100%

Considerable
effort

Minimal
cooperation

Option B

£10 = 11%

Fix profit as
lump sum

Jointly address
prime cost

£90 target
achieved

Prime cost
£80 = 89%

Cooperation

Figure 34.1 Options to reduce price by 10%.

Incentives may either be pre-planned or reactive to specific needs arising during the project. We often find that the motivation of the integrated team dips after the euphoria of the initial partnering workshop. As an example of a pre-planned incentive, the core group may decide to maintain team morale by planning a team event, in the form of a team lunch or outing, before the dip is likely to occur. Core groups should recognise the diverse nature of the team and choose activities that are appropriate to all team members and which will reinforce the team culture.

Note that in order to ascertain whether or not an incentive is likely to produce the required result, the organisation or person offering the incentive needs to understand the value criteria of the intended beneficiary. An organisation should not impose its own value criteria on others but should seek to understand what motivates their partner. Examples of incentives that we have seen placed on organisations include:

- elimination of retentions, more frequent certification and faster turnround of payments. This does not only apply to client/constructor relationships but also to constructor/specialist arrangements
- advanced payment for long lead items
- joint publicity to enhance reputation of all parties.

Beyond a fair return, the simplest way to introduce a financial reward scheme is to share any savings identified by the team. This should be done in a fair and equitable proportion and the process should be agreed before the start of the project. Such calculations do not need to be complex. There are many complex formulae, diagrams and weighted incentive matrices for apportioning savings and other benefits. These may be appropriate for major infrastructure works and similar scale projects but, in all cases, the team should bear in mind the cost of setting up, calculating and implementing a reward scheme. One that takes ten resource days at £300 per day to administer a saving of £2000 is not effective use of the integrated team's skills and resources.

As soon as we mention *sharing rewards* a shudder often runs round the public sector, accompanied by statements such as, 'There's no benefit to us in sharing savings' or 'We can't share – it's public money'. It is important that all members of the integrated team are aware that some clients will be unable to make a shared saving available as they may be constrained by public sector rules. In this case, the team should identify an alternative approach which will benefit all parties, such as reinvesting the saving in extra work that provides added value for the client and, at the same time, additional margin for the design and construction team. However, clients should note that if they will not share rewards or find alternative solutions to benefit their partners, there may be little or no incentive on the partners to notify the client of any savings they do identify.

> One partnering team that we worked with identified £5000 underspend at the end of the project. Rather than split the cash between the partners, the core group decided to provide additional paving and clothes lines for the residents, adding value through enhanced satisfaction of need for the residents and the client, whilst adding value through increased turnover and profitability for the constructor.

There should be a direct link between a reward and the outcome of the project. Potential added value factors that could be taken into account in determining the extent of a reward scheme are:

- delivery of the project under the budgeted figure, where this is of benefit to the client (note that some clients may not gain from an underspend as they may, for example, have to repay grants)
- project delivered early where this results in, for example, benefits of unanticipated early trading (note that some clients may not gain from an early delivery as they may incur early unbudgeted costs)
- zero defects leading to full utilisation of the asset (however, there is an argument that zero defects is what has been contracted, so no reward is due)
- reduced operating and whole life costs.

Note that the reward should not be for delivery to the contracted brief, but for exceeding client or team expectations. Core groups will need to consider in advance whether a reward is paid if, for example, the project delivers under budget (for which a reward may be appropriate) but late (in which case it is not).

Rewards should, generally, not be directed to a single team member or organisation but to the integrated team as a whole. In this way, the individuals and organisations will be incentivised to pull together as the excellence of individual stars is secondary to the efficiency of the integrated team. However, we have found that, for individuals, a simple 'thank you' can be a strong reward or motivating factor. This can be even more effective if it is backed up by a letter from a member of the core group.

Incentives and rewards are inextricably linked with risk. Early risk identification identifies ways of dealing with problems and helps teams deal more effectively, and therefore more profitably, with the project. We have identified, in the chapter on risk management, the need for the team to get together early in the project to identify the key risks to delivering the project on time, within budget and to the required quality to meet the needs of all the project participants. Not only is this an important stage in developing the partnering team's approach to the project but it may be a contractual commitment under partnering contracts. Some risks will be owned by the client, some by the constructor and some may be covered by a joint project risk pot. Where a risk has been taken by one party alone, it may be inequitable to expect that party to share the saving if the risk does not occur. Project risk pots, on the other hand, may provide the opportunity for reward share at the end of the project.

Having identified and planned for the risks, there remains the question of how the team will deal with any unused project risk pot at the end of the project. The simple answer to this is to be open and agree the procedures in advance. The following questions need to be addressed in setting any procedures:

- is the client prepared or able to share any remaining risk pot with the team?
- is the constructor prepared to share any remaining risk pot with the client?

If the answer to either of the above questions is, 'no', the team has an issue which requires resolution. However, if the answer to both is, 'yes', the core group should define at the start of the project what proportion goes to each member of the team and identify how and when the fund is transferred.

The core group has a major role to play in ensuring that incentives and rewards are in accordance with sector rules, appropriate to the project and motivate team members. Thus, one of the earliest actions of the core group must be to, '...agree such incentives...as may be appropriate to encourage partnering team members to maximise their efforts...for the benefit of the project' (Association of Consultant Architects & Trowers & Hamlin, 2000). The core group should, at the same time, develop and agree the process to be followed in sharing any added value delivered by a partnering or integrated team that delivers a project exceeding expectations. Appropriate incentivisation and reward will motivate the integrated team to pull together and go the extra mile to deliver added value for all.

35 Partnering Contracts

Since the publication of Sir Michael Latham's report, *Constructing the team* (Latham, 1994), an increasing proportion of the UK construction industry has sought to work in a partnering and integrated team-working approach to projects. Latham's principles of a modern construction contract included:

1. duties of fairness, mutual cooperation, teamwork, and shared financial motivation
2. a wholly integrated package of documents clearly defining and separating roles and duties and suitable for all projects and any procurement route
3. easily comprehensible language
4. agreed and appropriate allocation of risks for each project
5. avoiding changes but, where these occur, pricing them in advance
6. flexible provision for swift payment throughout the supply chain with clear and secure entitlements to payment and compensation for late payment
7. mechanisms for the avoidance of conflict and for speedy dispute resolution
8. incentives for exceptional performance.

Boosted by the publication of *Rethinking construction* (Egan, 1998), many extremely successful partnering relationships were set up with both public and private sector clients utilising pre-existing contracts such as the JCT, ICE and GC Works 1 forms with or without the use of practice notes, options and addenda.

The development of partnering or team-specific contracts such as the Engineering and Construction Contract (ICE, 1995), PPC2000 (Association of Consultant Architects Ltd & Trowers & Hamlins, 2000), the Be Collaborative Contract (Be, 2004) and the Public Sector Partnering Contract (Knowles & Wills, 2004) seek to put into contractual terms the successful processes and experiences of partnering teams, making the contracts readily usable by those who accept the need for a contemporary, cooperative approach to construction procurement.

There is still a body of opinion that supports the use of amended traditional forms of contract. It has clearly been possible to partner successfully without a partnering-specific contract, as many successful partnering arrangements were in place before the new forms were developed. However, we would generally recommend that teams now move towards partnering-specific forms. Using terminology specific to one or other of the partnering contracts is unavoidable but we would point out that, in using such terminology, we are not recommending one contract over another. The decision on whether to use a partnering-specific form of contract and, if so, which partnering-specific form to use, must be that of the integrated team.

A perceived advantage of traditional contracts is that everyone understands them and they can be signed and put away in a drawer until something goes wrong. However, the act of pulling the contract out of the drawer, after an issue has arisen, is a signal that the parties are looking to establish their rights and remedies rather than focusing on resolving the issue.

A disadvantage of traditional contracts is that they are single party contracts – one-on-one. This means that each project will involve multiple parties appointed on separate contracts with a variety of separate terms and conditions, with the contracted parties owing little or no allegiance to the project or to any other member of the team except the one with whom they are contracted. This may not only be difficult to manage but may also lead to the partners pulling in different directions to the detriment of the project.

The principle of partnering contracts is to address the issue of team focus on the success of the project by all partnering organisations signing up to the same contract. Most partnering contracts make it a condition that the partners work together for the benefit of the project and for the benefit of all parties using expressions such as mutual objectives, trust, good faith, fairness, cooperation, collaboration and respect.

An advantage of contracts that are drawn up specifically with partnering and integrated teamworking in mind is that the team does not have to find ways around the contract if they want to work cooperatively. The adjustment of contract rates through an open book approach is theoretically not possible with many traditional forms without a great deal of effort and ingenuity on the part of the partners. Partnering contracts facilitate such an approach if it can be proved to deliver better value.

Some of the specific requirements that differentiate, for example, PPC2000 (Association of Consultant Architects Ltd & Trowers & Hamlins, 2000) from traditional contracts include:

- a requirement to, '. . . work together and individually in the spirit of trust, fairness and mutual cooperation . . .'
- the formation of a core group of senior representatives of all partners, meeting on a regular basis with contractual responsibility to ensure that the partnering ethos is maintained
- an early partnering programme from the client representative followed by a project programme from the contractor before the project commences
- the requirement to, '. . . work together to . . . analyse and manage risks in the most effective ways . . .'
- provision to share value enhancements arising from joint work on value management
- processes for early identification and communication of problems
- identification and setting of KPIs, targets and measurable continuous improvement.

These are all aspects that may require major cultural shift on the part of organisations and team members using partnering contracts for the first time. The joint training that may be required to bring about this culture change will involve effort and will cost money and resource time to set up and facilitate. Having identified the cost and resource time required, some organisations decide to save money and effort by not following this programme through but our experience is that this cost-focused, rather than value-focused, approach is likely to add no value to the project. Saving money is not necessarily the same as adding value and our experience is that the earlier and the more committed the joint input to training and teamworking, the greater the added value to the project and to all partners.

The commitment to regular core group meetings may be another culture change for senior members of the partnering organisations. Depending on the form of contract, the role of the core group may be slightly different but the core group will probably be the highest level of issue resolution and decision making within the project team before resorting to alternative dispute resolution procedures such as mediation. Regular meetings are therefore essential to ensure that the core group members understand each other's ways of working so that the few issues that do escalate to their level are resolved jointly, speedily and effectively. Since the core group may also be responsible for setting incentives and reward, agreeing key performance indicators and driving the partnering ethos, the resource implications of committing to membership of the core group should not be underestimated.

Adhering to all the specific requirements of a partnering contract appears to require considerably more input than traditional contracts. However, our experience is that this is not the case when measured over the duration of the project. It is true that more input may be required at the beginning of the project but this reduces the effort required in later stages. Much of the work that is explicit in partnering contracts is undertaken implicitly by various team members in all contracts, whether partnering or not. This work on non-partnering contracts is frequently undertaken at risk of not being paid. Traditional contracts do not come into play in the early project stages and therefore do not specifically recognise or reimburse the resource allocated, for example, to joint design and tendering processes.

Nevertheless, all team members should be aware of the costs of operating in accordance with the formal requirements of a partnering contract and include these as part of the budgeting process for the project. An £8million project, six months in design and fifteen months on site will probably have an initial partnering workshop (PSPC calls for a two-day initial workshop), two workshops to initiate the value and risk management processes, six quarterly continuous improvement workshops (which may include team events) and a post-project review – a total of ten workshops. The cash outgoings can be taken as being in the order of £2500 plus VAT for each workshop (including the fee for the facilitator and the cost of hiring a venue) so the team should allow a sum in the order of £25k in their budget. It is our experience that effectively structured and facilitated workshops cut out many of the ad hoc and one-on-one meetings that are a feature of traditional relationships and save resource time as well as ensuring full team understanding and a common sense of purpose.

In our opinion, the benefits of a well-structured partnering contract outweigh amended traditional forms in a partnering and integrated teamworking environment. However, for each project, the integrated team must make their own choice and, to do this effectively, they should use the experience within the team and, perhaps through joint training, become as familiar with partnering forms of contract as they are with traditional forms.

Amongst other issues, the team acknowledged that the available floorspace was exceeded by the space requirements and wishes of the users. They developed a set of comments and actions which included:

- open plan – there is a need for flexible space use and for access to areas of the building whilst maintaining security and a means of safe access and egress for all
- the users and the design group will meet regularly to develop acceptable solutions
- the users will review their perceptions of who would be located on which floor and identify who would join the proposed meetings with the design team
- the users will meet over the next week to finalise the area allocations
- the initial scheme will be drawn up and circulated to the team by the architect within three weeks of the workshop.

In reviewing the workshop, Dorothy Hague highlighted the importance of early input from interested parties, commenting that, '. . . the areas you covered were crucial to the project and extremely valuable to the participants. You have certainly highlighted the hard work necessary to achieve the University's aims and it's best that we know that up front'. Two specific crucial aspects that Dorothy identified as helping to direct the process were identifying the core group members and establishing key dates over the first three months, which highlighted the urgency and helped motivate all parties.

The inclusion of procurement, finance or auditors in early team workshops (such as value and risk management) can assist with the quantification of added value. We have already shown, in the chapter on continuous improvement, how some *soft* benefits can be converted into cash equivalents to prove added value. This effort will be to no avail if the interested financial parties refuse to accept such calculations after the event – their agreement and commitment is required up front.

Interested parties can also include the wider community who may be impacted by a specific project. The following report was drawn up by a cross-organisational group in a continuous improvement review for The Oxfordshire Rural Housing Partnership and their constructor partner Leadbitter.

> The launch of the Oxfordshire Rural Housing Partnership achieved national coverage in the housing press less than a year ago. The momentum has continued at a local level with a hugely successful poster competition to raise awareness of site safety with schoolchildren close to the first partnership scheme to go on site, in Witney. A community development day is to be held later this month to engage the wider community and identify priorities for the use of a Community Fund jointly funded within the partnership.

The involvement of interested parties from an early stage of the project will help to build common understanding across all disciplines of the integrated team. Whilst continuity of personnel is desirable throughout the project, it is likely that new members will join. The core group should be aware that the relationships and knowledge built up over the preceding months within the team may make it difficult for a new member to break in. New members may include:

- representatives of organisations that have recently been appointed to the team (e.g. specialist constructors)
- team members representing new roles within established organisations (e.g. a post-contract quantity surveyor taking over from a pre-contract colleague)
- new members of established organisations (not only new to the project but new to their own organisation).

Organisations and integrated teams will have developed their own TLAs (three letter acronyms) and other jargon and existing team members may not always be conscious of the way in which these abbreviations exclude others. We know of one organisation, for example, that has a nine page guide to company acronyms.

We recommend that the core group should ensure that there is a partnering and integrated teamworking induction pack. The project manager or a core group member should spend a half hour with each new joiner on the day they join the team, discussing and detailing the integrated team approach to the project with them. Note that we have not said that the new team member should find this information on the project website or be told to read the partnering file inside or outside office hours. The proposal for a half hour induction may be greeted with concern by some but there is nothing better than face-to-face communication to ensure comprehension. Time devoted to a structured induction will be time well spent. Consider how many half hours will be wasted if new members fail to understand and apply the team ethos or feel unable to make an active contribution to a workshop because they have not been able to feel part of the team.

The induction should enable the new joiner to feel part of the team and contribute to the project from their first day. The induction pack should include:

- a summary of partnering and integrated teamworking as it applies to the project
- the (project) partnering charter
- the issue resolution process
- the client's value criteria
- KPIs and improvement targets
- explanations of project acronyms and jargon
- names and contact details (phone, fax, mobile, pager and email) of members of the partnering team including identification of core group members or partnering champions
- executive summaries of value and risk management and partnering workshops
- up-to-date partnering and/or project timetables
- schedule of all core group and partnering team meetings, workshops and social events
- information on where or from whom further information can be obtained.

The proactive involvement of interested parties throughout a construction project enables integrated teams to add value by understanding and responding to the needs of occupiers and operators and managing their expectations. In order to achieve this added value, it is critical that users, designers, constructors and all other members of the integrated team commit to meeting and discussing their needs and resources in a regular programme of partnering and integrated team workshops and events. Some team members may be reluctant to commit resource to such a programme. However, this commitment to integrated teamworking should reduce the volume of post-completion user problems and complaints. The result is not only increased user satisfaction, but also reduced post-completion resource waste in team members fielding complaints and dealing with rework. Win–win.

37 Avoiding the Pitfalls of Partnering

Partnering and integrated teamworking has been proved to deliver substantial benefits to organisations that commit to its implementation. However, there are partnering relationships that deliver sub-optimal performance or fail.

We have identified seven behaviours – the pitfalls of partnering – that characterise these less successful relationships. We have used the acronym PITFALL to identify these behaviours:

Personalities not combining as a team
I not 'we'
Traditional attitudes and behaviours
Fear of change
Adversarial attitudes
Lack of learning
Lip-service

Partnering and integrated teamworking depends on a continuous, consistent, proactive, team approach. After the honeymoon period, there is a danger of individuals or sections of the team reverting to type. The core group must be aware of this and address the situation promptly.

We suggest that any necessary cultural change is driven by the core group of partnering champions. This team within a team will have the role of change agents. Whilst the initial drive is necessary, it is also important that the core group perseveres in implementing partnering, holding regular continuous improvement workshops, monitoring key performance indicators and publicising success. An occurrence of any one of the seven pitfalls should alert the core group to take remedial action.

PERSONALITIES

The pitfall is that individuals (management and operatives) are unwilling or unable to fit into an integrated team. This can undermine and threaten the success of the entire relationship.

In order to avoid the pitfall, the core group must be aware of and address any behaviours that are detrimental to the interests of the integrated team. In the event of a personality clash within the team, the core group should identify whether the behaviour is:

1. an interface problem with another team member, in which case the core group should bring the two (or more) team members together, identify the issue and help the individuals to identify ways to move forward for the benefit of the project or the team.

or

2. an issue with the individual, in which case the core group should discuss the situation with the individual and consider options:
   a) retrain the individual
   b) allow the individual to remove him/herself from the partnering team
   c) in extreme cases, remove the individual from the partnering team.

In adopting a proactive response to non-partnering personalities, the core group will ensure that the team's partnering ethos is not undermined.

'I' NOT 'WE'

The pitfall is that individuals fail to work as part of an integrated team. They work for themselves or for their own organisation rather than for the team as a whole. They are not prepared to make any sacrifices or understand another's point of view, thinking only of their own self-interest. This fosters confrontation, endangers relationships, breaks trust and lowers morale.

In order to avoid the pitfall, the core group must actively promote the integrated team approach encouraging all team members to focus on the successful delivery of the project and thus look after their own and their partners' interests jointly. The core group should review the original partnering charter with the team and re-publicise this in order that the mutual objectives stated in the charter permeate the thinking and behaviour of all.

The team should be aware of non-partnering language and address this evidence of traditional or adversarial behaviour by positively acknowledging when partnering language is used (see Fig. 37.1).

Moreover, the core group should focus on the team's successes, especially identifying and publicising successes that have come about as a result of cooperation between team members.

TRADITIONAL APPROACH

A focus on lowest price rather than best value can blind the team to the benefits of integrated teamworking. Some team members may find it very difficult to banish the mindset of lowest price.

In order to avoid this pitfall, the core group should provide training which will need to be supplemented by continuous reinforcement of

Non-partnering language	Partnering language
• I	• We
• Them	• Us
• My project	• Our project
• They are not delivering	• We are not delivering
• Why did you do that?	• Where do we go from here?
• Why do you want it?	• How do you want it?
• The other side	• Our partners
• They	• The team
...plus...	...plus...
• Hidden agendas	• Straight talking
• Working against	• Working with
• Independence	• Interdependence

Figure 37.1 Non-partnering and partnering language.

the concept of best value. They should develop an integrated team focus on defining, identifying and delivering better value rather than lower price.

Joint workshops, open communication and a commitment to mutual objectives will ensure that a best value project is delivered successfully, meeting the objectives of the team and of individual partners, bringing into the team the previous advocates of the traditional approach.

FEAR OF CHANGE

Fear of change can undermine efforts to implement partnering and integrated teamworking.

Change is constant and seems daunting to individuals. Resistance to change is common and those who resist need to see a path to clear goals. It is critical to establish good two-way communication to identify, isolate and address fears.

In order to avoid this pitfall, the core group should listen to and understand the fears and concerns of individual team members. Members of the core group should be honest and open about the impact of the change and address individuals' concerns as far as possible within the context of an integrated team.

Whilst fears must be recognised and addressed there is a real danger that openly-expressed reservations will turn into self-fulfilling prophesies (e.g. 'We've seen it all before; nothing will change.'). Team members must be careful not to undermine the process by negative and ill-judged comment.

Fear is frequently a result of gossip, innuendo and false assumptions – the unknown future. In some cases this may be exacerbated by deliberate undermining of the initiative by some individuals. This can be overcome by good communication of the organisations' strategic objectives and the steps that will be taken to deliver them. Factual information will discourage rumour.

The core group should target the delivery and quantification of substantial value enhancements within a twelve month period. The team should develop ways of evaluating improved service, customer satisfaction and other efficiencies throughout the supply team.

Training in integrated teamworking and partnering and in handling the effects of change will assist team members in reducing their fear of the unknown. Such training will help build the team and help individuals buy into the initiative.

ADVERSARIAL ATTITUDES WITHIN THE TEAM

The pitfall is that individuals or organisations in the team adopt an adversarial attitude towards others both within their own organisation and within the integrated team. Even within a team whose objectives are clearly aligned, an individual can adopt an adversarial attitude in expressing their point of view. They may believe that their answer is the correct answer for the team and are unwilling to listen to alternatives. They talk too much, shout down others and don't value or respect the opinions of other team members. Similar issues arise through the use of aggressive non-verbal communication such as finger wagging, scowling and even banging the table.

In order to avoid this pitfall, the management of meetings should be firm, fair and effective. All team members must be allowed their input where this is valid to the item under discussion. The facilitator (or meeting manager) should set ground rules for meetings which could include:

- ❑ set an agenda with clear objectives and appropriate time slots for each topic
- ❑ keep to the agenda and to time
- ❑ identify what is and what is not within the remit of the meeting
- ❑ encourage discussion, discourage argument
- ❑ focus on fact, not on hypothesis by drawing out actual case studies/information
- ❑ focus on the problem, not on the person
- ❑ allow time for each team member to contribute but not excessively
- ❑ to avoid one individual hogging the meeting: break eye contact and direct a question or comment to another person
- ❑ write action points on a flip chart so all can see what has been agreed.

Good management of team meetings will promote cooperative working and develop mutual respect so all views are valued, even if they differ.

LACK OF LEARNING

The pitfall is in attempting to build an integrated team without training and without feedback from previous experience. A lack of training may lead to different perspectives and may make it difficult to focus the team on common goals, slowing the development of the integrated team to the detriment of the whole.

In order to avoid this pitfall, the team should implement joint training initiatives. It is critical to the success of integrated teamworking that all individuals involved have a common understanding of the principles and practices involved.

Joint training at the start of a relationship will help to:

❏ drive a common culture
❏ build the integrated team ethos
❏ reduce the per capita costs of training.

Feedback of successes and opportunities during or at the end of projects from the integrated team is critical to driving continuous improvement. The team should develop a structured process for feedback and knowledge management within and across all of the organisations to drive the benefit of the learning curve.

LIP-SERVICE

The pitfall is 'ticking the box' to secure funding, followed by a lack of commitment of financial or human resources. This will result in a failure to build the integrated team and, consequently, the delivery of less than optimal value for any or all of the partnering organisations.

In order to avoid the pitfall, each organisation should consider whether partnering or integrated teamworking is appropriate for them. Partnering may not be the most appropriate procurement route as some projects are more suited to a commodity buying procurement route and some organisations are culturally unsuited to a cooperative working environment.

In order to combat the culture of lip-service to partnering and to drive continuous improvement in value and team satisfaction, those who are committed to a partnering and integrated teamworking approach should adopt a structured process as follows:

- select team members on value-based criteria
- hold an initial partnering workshop to build the integrated team, setting out a statement of mutual objectives (charter), developing a robust issue resolution process and targeting continuous improvement
- set dates for continuous improvement workshops for the whole programme (including a post-project review) involving specialists, operatives and interested parties during the project and do not cancel them. Within each of these workshops the team should identify and set a limited number of clear actions to add value.

Between workshops:

- the core group should track actions on a regular basis and offer such assistance and resource as may be necessary to complete the action by the stated time
- the whole team should support the development and delivery against the actions, led by those who were assigned the action
- the core group should proactively support and provide resources to those carrying out the actions
- those nominated to lead the actions should report on the actual value enhancement at the first available opportunity
- individuals should work as integrated team members.

Finally, but most critically, the team must publicise their successes because many potentially excellent partnering and integrated teams have fallen apart because the team failed to supply decision makers with quantified evidence of success.

The steps required to avoid the pitfalls of partnering require the investment of considerable pre-project effort and resources from all who intend to partner. This investment is necessary in order to nurture a culture of trust and openness in which partnering can flourish. The return on investment will be significant – measured in terms of the delivery of better and faster projects with enhanced value for all involved.

38 Icebreakers and Team Building Exercises

In the course of facilitating more than 400 team workshops we have adopted, adapted and developed a variety of exercises to help the teams focus on specific aspects of the workshop or the relationship.

We work with some team leaders who specifically ask us to prepare exercises for workshops and others who say that their team only wants to concentrate on the technical issues in hand. Whilst we acknowledge a team's wish to concentrate on technical issues, there is considerable added value to be gained from team interaction in structured exercises. Any team exercise should feature:

- a learning point which is relevant and can be applied to the project
- time to review the learning point
- simple rules
- a clear explanation of the task from the facilitator and time for questions
- fun (we learn better when we are having fun)
- inclusion of all members of the team
- no embarrassment
- appropriate duration plus up to five minutes to review the learning point
- minimal setting up and clearing away
- low cost.

Facilitators may add one of the following icebreakers to the introductions in a workshop (in addition to the team members introducing themselves and explaining their role in the project):

- ❑ identify the number of years experience and add up the total for the team or their group
- ❑ say what their hobby is
- ❑ say what they are planning to do to relax next weekend
- ❑ give a title to their group by identifying a common theme amongst the members
- ❑ appoint a representative from each group to introduce their group colleagues after they have had time to introduce themselves to each other.

In the following team based exercises, we have highlighted the key learning point(s) and given an indication of the duration of the exercise.

COMMUNICATION – I DIDN'T SAY YOU WERE STUPID

A great proportion of meaning is in the way that the words are said. Even 'thank you' can take on different meanings depending on the intonation of delivery – for example, sarcasm or praise. We have set out below an exercise that can be used to demonstrate to partnering teams how a simple six-word sentence can take on multiple meanings depending on emphasis or intonation.

- ❑ On a flip chart write out the words 'I didn't say you were stupid' with each word on a separate line and without any punctuation.
- ❑ Sit the team in pairs facing each other and, within each pairing, ask them to decide who will be partner A and who will be partner B.
- ❑ Ask all the partner As to say the six words on the flip chart to the partner Bs; tell them to emphasise the first word and tell both partners to maintain eye contact throughout so that the facial expression of both partners is recognised.
- ❑ After the laughter has subsided, ask all the partner Bs to say the six words on the flip chart to the partner As, telling them to emphasise the second word and to maintain eye contact throughout.
- ❑ Repeat the previous stages, alternating the delivery between partner A and partner B and moving down the chart, emphasising a different word each time.
- ❑ Review the exercise with the team.

The learning point from this exercise is that there is more meaning in the intonation than in the written word. The exercise highlights the propensity for misunderstanding in the simple written word, especially emails, where there is no intonation to provide clues.
Time required – less than five minutes.

PARTNERING AND COOPERATION – RED–BLUE

Probably the best known partnering exercise is a development of the 'The Prisoners' Dilemma'. Many partnering teams will know this as 'Red–Blue'. The rules below should be shared with the participants but the notes in brackets should be removed as these are notes for the facilitator.

The team is divided into two groups – A and B. The objective of the exercise is to obtain the highest possible score. Each player deposits a cash stake with the facilitator at the beginning of the exercise. (The team members' focus on the exercise will be greater for a higher stake. We normally pitch the stake at about £3 but this should be at the discretion of the facilitator.)

Each group begins with zero points and will nominate two officials to the facilitator before the first round commences – a runner and a negotiator. If both groups end the game with a positive score the stakes are refunded. Note that zero is not a positive score. The facilitator will give each group eight cards, one for each round, on which the groups will identify whether they want to play red or blue.

In each round, the group will review the scoring matrix below and decide whether to play red or blue. They will then send the runner with the card indicating their decision to the facilitator.

At the end of rounds 3** and 6** (see Fig. 38.1) each group may indicate to the facilitator that they want to talk to or negotiate with the other group. Only if both groups want to talk is a meeting set up between the two negotiators. The negotiator must be briefed by their group before the meeting.

Points will accrue according to the following rules. The facilitator will keep score (although groups may also be provided with a scoring table).

If Group A plays BLUE and Group B plays BLUE then A scores −10 and B scores −10
If Group A plays BLUE and Group B plays RED then A scores +20 and B scores −20
If Group A plays RED and Group B plays RED then A scores +10 and B scores +10
If Group A plays RED and Group B plays BLUE then A scores −20 and B scores +20

In rounds 1 to 5 the scores are as above.
In rounds 6 to 8 inclusive the scores are doubled.

The first learning point from this exercise is that terminology used is important. In this case, groups make up a team and the groups should work together for the benefit of the team if they are to obtain the required result. The team should learn that the best result (the highest score) comes from the groups identifying that the cooperative red–red play is the only one that produces a positive score for the team in each round. Any pairing including the *selfish* blue option results in a loss to the team (blue–blue) or one group gaining at the expense of the other (blue–red) with no overall gain to the team. Note

Round	A's card	Score	Total	B's card	Score	Total
1						
2						
3**						
4						
5						
From this round all scores are doubled						
6**						
7						
8						
		Final score			Final Score	

Figure 38.1 Red–Blue scoring matrix.

that the highest possible (combined) team score is 220 points from both groups playing red all through the exercise. The team should also learn that each group trusting the other (and keeping to their negotiated word) is essential to achieving the maximum potential of the team.
Time required – between 40 and 60 minutes.

INTERDEPENDENCE – COOPERATE

In addition to using Interplace, the computer expert system developed by Belbin Associates (www.belbin.com) we regularly use their team role exercises. Cooperate uses three action-oriented exercises to illustrate and overcome common teamworking problems. We won't spoil the fun but simply point out that they all need good team cooperation.

There are different learning points for each of these exercises. Team Write calls for team harmonisation, Team Build calls for self-sacrifice and Team Rescue calls for good communication. Time required – around 30 to 40 minutes for each exercise.

BELBIN TEAM ROLES – CONTRIBUTE

In Contribute, tasks are assigned to team members according to their Belbin Team Role. Each team member must ensure completion of their task in order to contribute to the success of the team within the one-hour playing period.

The learning point from this exercise is that team members should practise and develop their preferred non-technical team roles in order to contribute to overall team success.
Time required – one hour plus debrief (around seventy five minutes in total).

RISK MANAGEMENT – LEGO® KIT

The facilitator should provide an unopened box of a small Lego® kit and set the team a challenging target (say 20 minutes) for completion of the kit in accordance with the illustration on the box. However, before the team can open the box (equivalent to starting on site) they must develop a register of key risks. The risks are those that will prevent them from completing the kit (delivering the project) in the set time in accordance with the illustration on the box and to the satisfaction of the facilitator (client).

In the first stage, identification, the team should be given ten minutes to identify ten risks (event and consequence).

After identification, the team must assess separately the likelihood and impact of each risk on a scale of 1 (negligible) through 2 (low) and 3 (medium) to 4 (high). They should be given a further ten minutes for this.

For the four top scoring risks (the highest scores when likelihood and impact are multiplied together), the team should identify a risk management plan. They should identify what they should do to avoid the event and what they will do if the event occurs. An owner should be identified for each risk. Allow ten minutes for this stage.

Once the risk management plan has been drawn up, the facilitator should start the clock to begin construction (which begins with opening the box). In the course of the exercise, the facilitator should observe:

- ❑ Did the team adhere to their management plan?
- ❑ Did the team members adhere to their allocated tasks?
- ❑ Did risk owners continue to monitor their risks or did they fall back into their technical task (e.g. assembling)?
- ❑ Did the team use any language of blame or was the language positive and supportive?
- ❑ Was time monitored?
- ❑ A Lego® kit regularly has some parts which are not in the design. Were these used to add value to the final model or were they wasted?

The learning point from this exercise is to apply the four stages of risk management in a fun situation to reinforce learning. Note that this works best with teams of around seven members. In a larger workshop environment, the team could be split into groups, each with an identical kit and set of instructions.

Time required – 60 minutes including debrief.

ACTIVE LISTENING – AS WE UNDERSTAND IT

We developed this exercise in response to a comment that project teams are excellent at their technical job but may not always listen to each other's points of view. The exercise focuses on developing the skill of active listening.

Stage One
Group A moves to a separate area for up to five minutes to prepare a two minute presentation on their issue (problem). The headline issue will be written on a sheet and handed to the facilitator. Group A will choose a presenter to deliver and explain the issue to Group B.

Stage Two
Group A's presenter will present the issue for a maximum of two minutes to Group B. There will be no questions or discussion during the presentation. Members of Group B may not write any notes – they will concentrate on listening to the presentation, attempting to understand Group A's point of view.

Stage Three
Group B will retire for up to five minutes to discuss and review Group A's issue. They will prepare a two minute (maximum) presentation on Group A's issue to present back to them (for example, 'As we understand it, your issue is . . . '). The presentation will not be interrupted by Group A.

Stage Four
Group A will confirm that Group B understands the issue or, if this is not the case, will restart the process. When understanding is confirmed, the groups will jointly address the issue as an integrated team and seek resolution, adding value for all.

Figure 38.2 describes the process for Group B actively listening to Group A's issue and resolving this as an integrated team. The process can be repeated for Group A listening to Group B's issue.

The learning point from this exercise is that team members should take time to actively listen to each other in order to fully understand each other's issues and jointly resolve them.

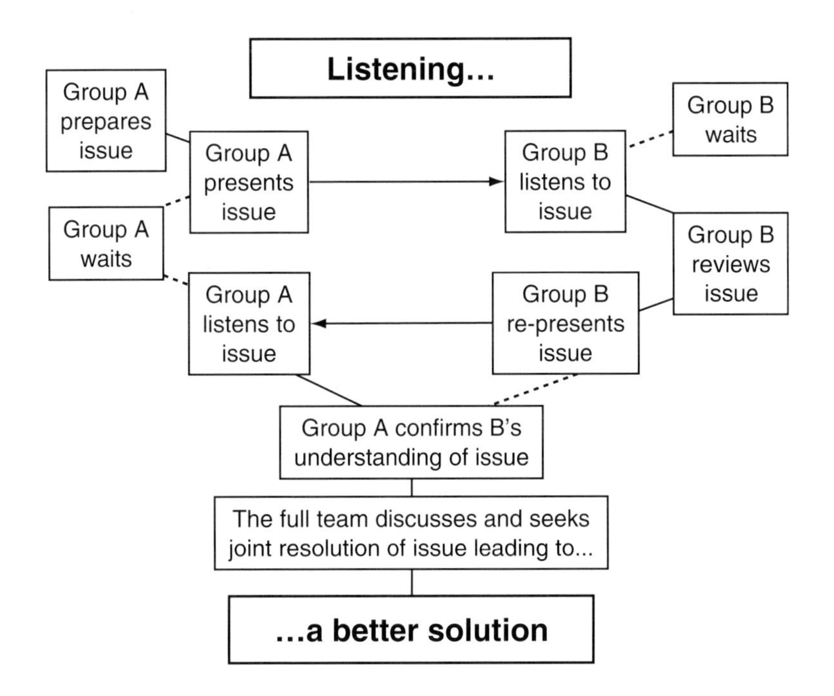

Figure 38.2 The listening exercise.

Time required – Stages One to Three will take around 15 minutes. If understanding is confirmed, the team will move directly to Stage Four and jointly resolve the issue. If understanding is not confirmed, the team will repeat Stages One to Three before moving to Stage Four.

SUSTAINABLE DEVELOPMENT QUIZ

This quiz of 20 questions has been taken from http://www.sustainable-development.gov.uk/sustainable/quiz/quiz.htm (DEFRA, 2004) and reproduced here with their permission. The leaflet *Sustainable development in your pocket* is available from Defra Publications, Admail 6000, London, SW1A 2XX. The quiz can be used in full or in two blocks of ten questions depending on the time available in the workshop. It can also be conducted in full team session or in groups.

1. By what percentage have UK emissions of greenhouse gases changed since 1990?

A. +16%; B. −14%; C. +9%; D. −19%

Answer: B – emissions fell from 208 million tonnes (carbon equivalent) in 1990 to 170 million tonnes in 2003.

2. Roughly what percentage of household waste in England and Wales is recycled or composted?

A. 35%; B. 25%; C. 15%; D. 5%

Answer: C – in 2002/3 520 kg of household waste was collected per person, of which 74 kg (14.2%) was recycled or composted.

3. Which of the following is roughly equivalent in weight to an adult male Indian elephant?

A. The average amount of household, commercial and industrial waste produced each year per person in the UK; B. The average amount of CO_2 emissions attributable to domestic energy use per person in the UK; C. The amount of remaining discovered and recoverable oil reserves within the UK per person?; D. An adult male African elephant.

Answer: A – an estimated 220 million tonnes of household, commercial and industrial waste was generated in 2000/1, equating to about 3.8 tonnes per person.

4. What percentage of people in England said that they regularly use low-energy light bulbs?

A. 9%; B. 16%; C. 31%; D. 48%

Answer: C – respondents were prompted with a selection of (environmental) actions and asked to what extent they did them.

5. What percentage of people in England regard their quality of life as fairly or very good?

A. 29%; B. 42%; C. 67%; D. 83%

Answer: D – when quality of life was defined in terms of how people feel overall, their standard of living, their surroundings, friendships and how they feel day to day, 27% of respondents rated their quality of life as very good and a further 56% fairly good.

6. What percentage of people in England said that they could easily access a local green space or local countryside without using a car or other transport?

A. 39%; B. 52%; C. 73%; D. 84%

Answer: D – respondents were also asked how often they used their local green space or local countryside. 73% had visited them in the last year, and 35% did so at least once a week.

7. What percentage of people in England said that they had heard of the term 'sustainable development' (before the World Summit in 2002)?

A. 22%; B. 34%; C. 46%; D. 57%

Answer: B – this had not changed between surveys in 1996/7 and 2001, and does not take account of whether they understood the term.

8. What percentage of children in Great Britain walk to school?

A. 81%; B. 64%; C. 44%; D. 31%

Answer: C – between 1985–6 and 2002 the proportion travelling to school by car doubled from 16% to 31%.

9. What percentage of electricity generated in the UK comes from renewable sources?

A. 3%; B. 8%; C. 13%; D. 18%

Answer: A – between 1990 and 2002 electricity generated by renewables, including hydro-power, increased by 60%.

10. What percentage of new homes are built on 'brownfield' (redeveloped) land?

A. 31%; B. 47%; C. 55%; D. 66%

Answer: D – increasing from 54% in 1990 and including conversions which account for about 3 percentage points.

11. What percentage of people of working age are in employment?

A. 52%; B. 66%; C. 75%; D. 82%

Answer: C – the percentage of working age people in work is about the same as it was in 1990.

12. What percentage of freight in Great Britain is transported by rail?

A. 2%; B. 8%; C. 18%; D. 25%

Answer: B – between 1970 and 2001, the proportion of freight moved by rail fell from 18% to 8%, though there has been a slight increase in recent years.

13. Approximately how many overseas flights were made by UK residents in 2002?

A. 12 million; B. 26 million; C. 39 million; D. 44 million

Answer: D – overseas flights by UK residents more than quadrupled between 1980 and 2002.

14. What factor is most often mentioned by people as affecting their quality of life?

A. Money; B. Health; C. Family and friends; D. Transportation

Answer: A – 48% of respondents mentioned money, 34% health, 7% family and friends, 13% transport. Respondents could mention more than one factor.

15. To what extent have CO_2 emissions from transport changed since 1970?

A. increased by less than a third; B. increased by less than two thirds; C. more than doubled; D. more than tripled.

Answer: C – transport emissions have increased by 130%, broadly in line with the increase in road traffic.

16. Based on CO_2 emissions per head of population, which of these rankings is correct (from highest emissions to lowest)?

A. Russia, United States, Australia, UK; B. United States, Australia, Russia, UK; C. United States, Russia, UK, Australia; D. Russia, United States, UK, Australia

Answer: B – in 1999 emissions per capita were as follows: US 19.9 tonnes, Australia 17.0 tonnes, Russia 10.2 tonnes, UK 9.0 tonnes.

17. To what extent have the number of deaths from circulatory diseases in England and Wales changed since 1970?

A. fallen by half; B. fallen by a quarter; C. stayed the same; D. doubled

Answer: A – in 2001 deaths per 100 000 were 47% of the rate in 1970.

18. According to the British Crime Survey by what percentage have vehicle crimes (theft of or from vehicles) changed since 1991?

A. increased by 30%; B. increased by 65%; C. decreased by 20%; D. decreased by 45%

Answer: D – in 1991, there were an estimated 3 845 000 vehicle crimes, by 2003/4 this had fallen to an estimated 2 121 000.

19. To what extent have farmland bird populations changed since the mid 1970s?

A. stayed roughly the same; B. doubled; C. fallen by half; D. fallen by a quarter

Answer: C – the index of farmland bird populations has nearly halved since its 1977 peak and has fallen by 18% since 1990. It has remained at about the same level over the last four years.

20. In real terms, to what extent has the cost of motoring changed since the 1970s?

A. increased by half; B. increased by a third; C. stayed roughly the same; D. decreased by a quarter

Answer: C – public transport costs rose by about 75% in real terms between 1974 and 2002. In contrast, the real cost of motoring has remained virtually unchanged, despite an increase in the real cost of fuel over the last decade.

The learning point from this exercise is to raise awareness of sustainability issues. If this exercise is carried out in groups, a further lesson is that the combined knowledge of the group is higher than that of any individual.

Time required should be in the order of 30 minutes.

References

Association of Consultant Architects Ltd & Trowers & Hamlins (2000) *The ACA Standard Form of Contract for Project Partnering PPC2000*. Association of Consultant Architects Ltd, Bromley.

Association of Consultant Architects Ltd & Trowers & Hamlins (2003) *Guide to the ACA Project Partnering Contracts PPC2000 and SPC2000*. Association of Consultant Architects Ltd, Bromley.

Be (2004) *The Be collaborative contract*. Collaborating for the built environment, Reading.

Belbin, R.M. (1981) *Management teams, why they succeed or fail*. Butterworth Heinemann, Oxford.

Bennett, J. & Jayes, S. (1995) *Trusting the team*. Centre for Strategic Studies in Construction, Reading.

Broome, J. (2002) *Procurement routes for partnering*. Thomas Telford, London.

Bruntdland, G. (ed) (1987) *Our common future*. Oxford University Press, Oxford.

Building (2002) Richard Hodkinson Consultancy quoted in *The New Building Housing supplement, February*. The Builder Group, London.

Byatt, I. (2001) *Delivering better services for citizens*. Department for Transport, Local Government and the Regions, London.

Chartered Institute of Public Finance and Accountancy (2003) *How to develop a Procurement Strategy*. Chartered Institute of Public Finance and Accountability, London.

Construction Confederation (1999) *Guide to waste reduction on construction sites*. Department of the Environment, Transport and the Regions, London.

Constructing Excellence (2003) *Demonstrations of sustainability*. Constructing Excellence, London.

Constructing Excellence (2004) *Construction industry key performance indicators*. Constructing Excellence, London.

Construction Industry Council (2002) *Guide to Project Team Partnering*, 2nd edn. Construction Industry Council, London.
Construction Industry Institute (undated) CII Research Summary 24–1 Cost-Trust relationships (US copyright).
Covey, S.R. (1989) *The seven habits of effective people*. Simon and Schuster, London.
De Bono, E. (1970) *Lateral thinking*. Penguin Books, London.
De Bono, E. (1998) *Simplicity*. Penguin Books, London.
DEFRA (2004) *Sustainable development in your pocket*. DEFRA Publications, London.
Egan, J. (1998) *Rethinking Construction*. Department of the Environment, Transport and the Regions, London.
Egan, J. (2002) *Accelerating Change*. Rethinking Construction, London.
Green, D. (2000) *Local Authority Guide to Rethinking Construction*. Local Government Task Force, Rethinking Construction, London.
Housing Forum (2002) *20 steps to encourage the use of Whole Life Costing*. Housing Forum, London.
ICE (1995) *The Engineering and Construction Contract*, 2nd edn. Thomas Telford, London.
Knowles, R. & Wills, M. (2004) *Public Sector Partnering Contract*. BLISS, Warrington.
Latham, M. (1993) *Trust and Money*. HMSO, London.
Latham, M. (1994) *Constructing the team*. HMSO, London.
Lundin, S.C., Paul, H. & Christensen, J. (2000) *Fish!* Hodder & Stoughton, London.
Marriott, J.W. (undated) *Management General Ezzay* from an excerpt at http://lists.webvalence.com/sites/ListeningLeader/Broadcast.D20040802.html
Mehrabian, A. (1981) *Silent Messages*. Wadsworth, Belmont California, USA.
MOD (2002) *Smarter Partnering*. Ministry of Defence, London.
ODPM (2003) *Best Value and Performance Improvement Circular 03/2003*. Office of the Deputy Prime Minister, London.
PSL (2004) *Partnering for profit*. April 2004 ed. Partnership Sourcing Ltd, London.
Rethinking Construction (2002) *Respect for people – reaching the standard*. Rethinking construction, London.
Strategic Forum for Construction (2003) *Integration toolkit*. www.strategicforum.org.uk/sfctoolkit2/home/home.html
Thomas, M. (2003) *COLA – the cross organisational learning approach for managing knowledge in partnering teams*. Hong Kong Institute of Value Management, 6th International Conference, Hong Kong.
Womack, J.P. & Jones, D.T. (1996) *Lean thinking*. Simon and Schuster, London.

WEBSITES

Belbin Associates: www.belbin.com
Construction Study Centre: www.constructionstudycentre.co.uk
Construction Industry Institute: http://construction-institute.org
Construction Clients Charter: http://www.constructionsuccess.org
Constructing Excellence: www.constructingexcellence.org.uk and www.kpizone.com
Considerate Constructors Scheme: www.considerateconstructorsscheme.org.uk
COLA application guide: http://is.lse.ac.uk/b-hive/COLA_application_guide.pdf
Dr Albert Mehrabian: www.kaaj.com/psych
Dr Tom Sant: www.santcorp.com
Institute of Value Management: www.ivm.org
Integration Support Network: www.integrationsupportnetwork.org.uk
Investors in People: www.iipuk.co.uk
Partnership Sourcing Ltd: www.pslcbi.com
Strategic Forum for Construction: www.strategicforum.org.uk
The Association of Consultant Architects: www.acarchitects.co.uk

Index